SpringerBriefs in Environment, Security, Development and Peace

Mediterranean Studies

Volume 8

Series Editor

Hans Günter Brauch

For further volumes:
http://www.springer.com/series/11792
http://www.afes-press-books.de/html/SpringerBriefs_ESDP_MeS.htm

Selim Kapur · Sabit Erşahin
Editors

Soil Security for Ecosystem Management

Mediterranean Soil Ecosystems 1

Springer

Editors
Selim Kapur
Departments of Soil Science and Archaeometry
Çukurova University
Adana
Turkey

Sabit Erşahin
Department of Forest Engineering
Çankırı Karatekin University
Çankiri
Turkey

ISSN 2193-3162 ISSN 2193-3170 (electronic)
ISBN 978-3-319-00698-7 ISBN 978-3-319-00699-4 (eBook)
DOI 10.1007/978-3-319-00699-4
Springer Cham Heidelberg New York Dordrecht London

Library of Congress Control Number: 2013944547

Cover photograph: A Mediterranean Soil Landscape on the Southern Coast of Anatolia, Ancient Kalantos (Kaledran) (Archive of Department of Archaeometry, University of Çukurova)

Printed on acid-free paper

Springer is part of Springer Science+Business Media (www.springer.com)

Mediterranean Soil Ecosystems (MSE)
Publication of the Soil Science Society of Turkey (SSST)

Editors: Selim Kapur and Sabit Erşahin

Managing Editor

Erhan Akça

International Advisory Board

Preface

Mediterranean Soil Ecosystems (MSE) is a peer-reviewed series of formal scientific publications of the *Soil Science Society of Turkey* (SSST). The SSST was established in 1964 and promotes research into soil science and related disciplines, as well as developing a consensus for soil protection and use to sensitize governments and the coming generations in Turkey.

Mediterranean Soil Ecosystems publishes original and review papers on the global and regional scales concerning theoretical and experimental studies and themes on soil ecosystems, soil and land degradation and desertification, agroecosystem management, carbon dynamics and management systems and ancient land use in the Mediterranean environment and context, palaeopedology and geopedology, and the changing soils and soils of tomorrow and their probable use under climate change scenarios. Special sections are devoted to current news from the *International Union of Soil Sciences*, the Soil Science Societies of the Mediterranean riparian countries, and the *Soil Science Society of Turkey* (SSST).

Mediterranean Soil Ecosystems also aims to improve communication and develop holistic integrated approaches among different disciplines seeking the sustainable management of the environment. Contributions are drawn from soil and earth sciences, agriculture, forestry, biology, botany, climatology, ecology, ecological economics, environmental sciences and engineering, environmental law, carbon policies, and information sciences related to environmental integrity. Mediterranean Soil Ecosystems publishes studies by academic researchers and professionals at universities, including those in business, government, research institutes, and public interest groups, presenting a wide range of approaches.

Adana, Turkey, March 2013

Selim Kapur
Sabit Erşahin

Contents

Contributors

Erhan Akça Technical Programmes, Adıyaman University, 02040 Adıyaman, Turkcy

Riccardo Biancalani Food and Agriculture Organization of the United Nations, Rome, Italy

Winfried E. H. Blum Department of Forest and Soil Sciences, University of Natural Resources and Applied Life Sciences (BOKU), Vienna, Austria, e-mail: winfried.blum@boku.ac.at

Pierluigi Bucelli Research for Agrobiology and Pedology, Florence, Italy

Edoardo Costantini Research for Agrobiology and Pedology, Florence, Italy

Selim Kapur Department of Soil Science and Archaeometry, University of Çukurova, Adana, Turkey

Ahmet R. Mermut Harran University, Şanlıurfa, Turkey; University of Saskatchewan, Saskatoon, SK, Canada

Luca Montanarella DG JRC, European Commission, Ispra, Italy

Freddy Nachtergaele Food and Agriculture Organization of the United Nations, Rome, Italy

Franco Previtali Department of Environmental and Earth Sciences, Geopedology, University of Milano-Bicocca, Piazza della Scienza 1, 20126 Milano, Italy, e-mail: franco.previtali@unimib.it

Lal Rattan School of Environment and Natural Resources, Ohio State University, 422B Kottman Hall, 2021 Coffey Rd, Columbus, OH 43210, USA, e-mail: lal.1@osu.edu

Alexander Tsatskin Zinman Institute of Archaeology, University of Haifa, 31905 Haifa, Israel, e-mail: tsatskin@research.haifa.ac.il

Pandi Zdruli International Centre for Advanced Mediterranean Agronomic Studies, Mediterranean Agronomic Institute of Bari (CIHEAM IAMB), Bari, Italy

Claudio Zucca Department of Territorial Engineering and Desertification Research Group, University of Sassari, Viale Italia 39, 07100 Sassari, Italy, e-mail: clzucca@uniss.it

Chapter 1
Managing Terrestrial Carbon in a Changing Climate

Rattan Lal

Abstract The threat of abrupt climate change by increase in atmospheric concentration of CO_2 and other greenhouse gases has enhanced the interest and urgency of identifying strategies for reducing and sequestering anthropogenic emissions. The latter are caused by land use conversion that began with the dawn of settled agriculture several millennia ago, and by fossil fuel combustion that began with the onset of the industrial revolution in about 1750. Emissions from land use conversion during the pre-industrial era until about 1850 are estimated at ~320 Pg. Since 1850, emissions from fossil fuel combustion are estimated at ~350 Pg and those from land use conversion at ~150 Pg. These and other anthropogenic activities have caused drastic perturbation of the global carbon cycle with increase in the atmospheric C pool and an attendant decrease in the pedologic, biotic, and geologic (fossil fuel) pools. Together, the pedologic pool (4,000 Pg to 3 m depth) and the biotic pool (620 Pg), called the terrestrial pool, is the third largest pool, after the oceanic (38,000 Pg) and the geologic (~5,000 Pg). The depletion of the terrestrial C pool has created a C sink capacity which can be filled by conversion to a restorative land use and adoption of recommended soil, plant, and animal management practices. The process of transfer of atmospheric CO_2 into the pedologic and biotic pools is called carbon sequestration. This natural process contrasts with that of the geoengineering techniques of *carbon capture and storage* (CCS) involving geologic and oceanic storage and mineral carbonation of CO_2 into calcite etc. The strategy of biosequestration, in addition to being cost-effective, has numerous ancillary benefits. It is a truly win–win option. Specifically, it improves soil quality, enhances agronomic productivity, and advances food security. Improvement in soil quality by C sequestration is related to generation and stabilization of micro-aggregates created through formation of organo-mineral complexes. The strategies of biosequestration involve development of a positive ecosystems C budget in soil by mulch farming, conservation

R. Lal (✉)
Carbon Management and Sequestration Center, The Ohio State University,
Columbus, OH 43210, USA
e-mail: lal.1@osu.edu

S. Kapur and S. Erşahin (eds.), *Soil Security for Ecosystem Management*,
SpringerBriefs in Environment, Security, Development and Peace 8,
DOI: 10.1007/978-3-319-00699-4_1,

agriculture, no-till systems, integrated nutrient management including biological N fixation and mycorrhizae use of amendments including biochar, and adoption of complex farming systems such as agroforestry. There is no silver bullet or panacea, and the choice of a practice/strategy depends on site-specific conditions.

Keywords Carbon sequestration · Geoengineering · Soil quality · Ecosystem services · Carbon capture and storage · Conservation agriculture · Soil structure

1.1 Introduction

Enrichment of the atmospheric concentration of CO_2 from 280 ppmv in the pre-industrial era to 390 ppmv in 2010, and the attendant increase in risks of *abrupt climate change* (ACC), have created an urgency to identity strategies of managing the *global carbon cycle* (GCC), and to limiting the increase in global temperature to 2 °C (Ramanathan and Xu 2010; UNFCCC 2009). Thus, establishing the cause-effect relationships for enrichment of atmospheric CO_2 is important to systemically reducing emissions and mitigating ACC. The importance of fossil fuel combustion since the onset of the industrial revolution during the Anthropocene (Crutzen 2002; McNeill 2000) is widely recognized. Global emissions from fossil fuel combustions increased dramatically during the second half of the twentieth century (Sterman 2008). Yet the role of land use change in emitting CO_2 and other *greenhouse gases* (GHGs) into the atmosphere cannot be overemphasized (Foley et al. 2005). In 1700, about 50 % of the terrestrial biosphere was wild, and most of the remainder (45 %) was in a semi-natural state. By 2000, most of the biosphere had been converted into "anthromes" consisting of croplands, grazing lands, plantations, and urban ecosystems and rural communities (Ellis et al. 2010). About 39 % of the earth's ice-free surface had been converted into agricultural and urban ecosystems, and an additional 37 % is embedded within or in close proximity to anthromes and is drastically influenced by anthropogenic processes. Transformation of natural/wilds to anthromes leads to depletion of the terrestrial C pool and emissions of CO_2 and other GHGs into the atmosphere. Depletion of the terrestrial C pool causes degradation of ecosystem services and functions. Management and restoration of the terrestrial C pool is essential to restoring the ecosystem functions, improving the environment, and mitigating ACC.

The objective of this chapter is to describe the relative contributions of land use change and fossil fuel combustion to the emission of CO_2 and other GHGs into the atmosphere, to identify processes and practices of C sequestration in the biosphere, and to explain the importance of soil C sequestration to *adapting to and mitigating* (ADAM) ACC, improving the environment, and advancing food security. Furthermore, processes of soil C sequestration are discussed in the context of enhancing permanence or *mean residence time* (MRT) and improving ecosystem services.

1.2 Relative Contributions of Fossil Fuel Combustion Versus Land Use Change

From 1850 to 1998, approximately 270 ± 30 Pg C had been emitted as CO_2 into the atmosphere from fossil fuel burning and cement production. In comparison, about 136 ± 55 Pg had been emitted as a result of land use change (IPCC 2000), of which 78 ± 12 Pg was from world soils. Another estimate showed that between 1750 and 2002, 292 Pg CO_2–C was contributed by fossil fuel combustion, and an additional 200 Pg C was projected to be emitted between 2003 and 2030 (Holdren 2008). Houghton (2007) estimated the total magnitude of anthropogenic emissions at 500 Pg since 1850, comprising 375 Pg from fossil fuel combustion, 100 Pg from land use change, and 25 Pg from cement production. Of these emissions, 150 Pg have been adsorbed by the oceans, 125 Pg by the terrestrial biosphere, and the remainder (225 Pg) by the atmosphere (Houghton 2007). The data in Table 1.1 show a progressive increase in CO_2–C emissions from fossil fuel combustion, beginning with merely 3 Tg (teragram = 10^{12} g = million metric tonnes) in 1750 to 8.7 Pg (petagram = 10^{15} g = billion metric tonnes) in 2008. According to these data, total emissions by fossil fuel combustion since 1750 are estimated at 350 Pg (Marland et al. 2007). At present, about 1.6 Pg C/year are contributed (~17 % of total) from land use change, involving primarily deforestation in the tropics. As late as the early 1950s, more CO_2–C emissions were contributed by land use (deforestation) than by fossil fuel combustion. Global average per capita CO_2 emissions have doubled from 0.65 Mg CO_2 in 1950 to 1.2 Mg in 1970, and have remained stable since then (Oelkers and Cole 2008).

Both fossil fuel combustion and land use change are driven by the increase in population (Table 1.2). The world population increased from merely 2 million in ~10,000 BCE to 188 million in 1 CE, 1 billion in about 1,800 CE, and 7 billion in 2011. The increase in human population resulted in an increase in cropland area from <5 Mha ~5,000 BCE to 300 Mha in 1,600 CE, 419 Mha in 1,800, 850 Mha

Table 1.1 Global estimates of fossil fuel emissions

Years	Total emissions (Tg C/year)
1750	3
1800	8
1850	54
1900	534
1950	1,630
1960	2,578
1970	4,075
1980	5,297
1990	6,096
2000	6,744
2008	8,700

Source Adapted from Marland and Rotty (1984), ORNL (2001), LeQuéré et al. (2010)

Table 1.2 Temporal changes in population, and total and per capita croplands and pastureland area

Time	Population 10^6	Cropland		Pastureland	
		Total (10^6 ha)	Per capita (ha)	Total (10^6 ha)	Per capita (ha)
10,000 BCE	2	0	0	0	0
5,000 BCE	18	4.8	0.24	0.4	0.02
1 CE	188	131	0.52	106	0.56
500	210	124	0.43	108	0.51
1,000	295	153	0.36	143	0.48
1,500	461	232	0.33	224	0.49
1,600	554	255	0.29	288	0.52
1,700	603	300	0.30	324	0.54
1,800	989	419	0.24	513	0.52
1,900	1,654	850	0.35	1,293	0.78
1,950	2,545	1,214	0.33	2,466	0.97
2,000	6,145	1,532	0.16	3,429	0.55

Source Adapted from Goldewijk et al. (2011)

in 1,900, and 1,500 Mha in 2,000. There was a similar increase in the area under grazing land/pasture land. The area under grazing/pasture land increased from 0.4 Mha ~5,000 BCE to 288 Mha in 1,600 CE, 513 Mha in 1,800, 1,293 Mha in 1900, and 3,429 Mha in 2,000 (Table 1.2).

Because of the drastic transformation of earth by humans since the transition of hunter/gatherer societies to a sedentary lifestyle and settled agriculture, some argue that the anthropogenic greenhouse era began thousands of years ago (Ruddiman 2003, 2006). Indeed, records show that the increase in CO_2 emissions began with the start of forest clearance 8,000 years ago and the increase in CH_4 with the onset of rice cultivation and the domestication of animals about 5,000 years ago. Ruddiman (2003) hypothesized that per annum rates of C release of CO_2 from land use in pre-industrial times may have been lower by an average factor of 10 or more. Even so, the cumulative emissions over millennia (for 8,000 years) could still be enormous. Total emissions by land use conversion of 480 Pg have been estimated as follows (Ruddiman 2003):

$$\text{(i) the pre-industrial era: 7800 years} \times 0.04\,\text{Pg C/year} = 320\,\text{Pg} \quad (1.1)$$

$$\text{(ii) the industrial era: 200 years} \times 0.8\,\text{Pg C/year} = 160\,\text{Pg} \quad (1.2)$$

Indeed, the greatest land clearance occurred during the last 200 years, with a total mean annual flux of 1.04 Pg C between 1850 and 2000, and as much as 2 Pg C between 1980 and 2000 (Table 1.3). Thus, cumulative C emitted from land use conversion has been estimated at 480 Pg over the last 8,000 years, equivalent to the enrichment of atmospheric concentration of CO_2 by ~120 ppm (4 Pg CO_2–C emission = 1 ppmv CO_2 concentration in the atmosphere) (Broecker 2007).

These statistics of CO_2–C emission from land use change (vs. fossil fuel emissions) for the last 8,000 years or more, tentative and crude as they may be, are

Table 1.3 Estimates of average annual flux of CO_2–C from land use change

Region	Annual emission (Pg C/year)		
	1850–2000	1980–1989	1990–1999
Tropics	0.65	1.93	2.20
Temperate	0.39	0.06	−0.02
Total	1.04	1.99	2.18

Source Adapted and recalculated from Houghten (2003)

important to the identification of strategies of C sequestration in the terrestrial biosphere. The fact that the terrestrial biosphere has lost as much as 480 Pg is a strong indication of its large C sink capacity. For this reason, recarbonization of the biosphere, enhancing the C pool in both soils and vegetation, is an important option to mitigating ACC.

1.3 The Global Carbon Cycle

The GCC involves principal C reservoirs and fluxes among them. The contemporary GCC involves five principal pools (Fig. 1.1). The largest pool is carbonate rocks (65 × 106 Pg), followed by ocean comprising 38,000 Pg C, mostly as inorganic C. The second largest pool is geologic, comprising fossil fuels, coal, oil, gas, shale, and peat, together estimated at ~5,000 Pg. The third largest pool is soil, containing ~4,000 Pg of *soil organic carbon* (SOC) and *soil inorganic carbon* (SIC) to 3 m depth. The fourth largest pool is the atmosphere containing ~800 Pg C. Thus, atmosphere merely contains 0.001 % of the C contained in the atmosphere–ocean–upper crust system (Oelkers and Cole 2008). The mass of the Earth's atmosphere is 5.14 × 1,018 kg (Trenbath et al. 1988); with a CO_2 concentration of 390 ppmv, total mass of CO_2 is about 3,000 Pg or equivalent to about 800 Pg C (Oelkers and Cole 2008). The smallest pool is biotic, estimated to contain 620 Pg C, comprising 560 Pg of live material and 60 Pg of detritus material. Combined together, the soil (4,000 Pg) and the biotic (620 Pg) pools, or the terrestrial pool, is estimated at 4,620 Pg, containing about 4.8 times more C than the atmospheric pool.

The GCC on a decadal scale from 1960 to 2008 is shown in Table 1.4. The cumulative annual sources of CO_2 were 4.6 Pg during the 1960s, 6.0 Pg during the 1970s, 7.0 Pg during the 1980s, 8.0 Pg during the 1990s, and 9.1 Pg during the 2000s (2000–2008), with total annual emissions of 9.9 Pg (8.7 from fossil fuel combustion and 1.2 Pg from land use change) in 2008. Of the total annual emissions, land use change contributed 32.6 % in the 1960s, 21.7 % during the 1970s, 21.4 % during the 1980s, 20.0 % during the 1990s, and 15.4 % during the 2000s (2000–2008), with a contribution of 12.1 % during 2008 (Table 1.4). Whereas the emissions from fossil fuel combustion have increased between 1960 and 2008 at the mean annual rate of 0.117 Pg C/year, those from land use

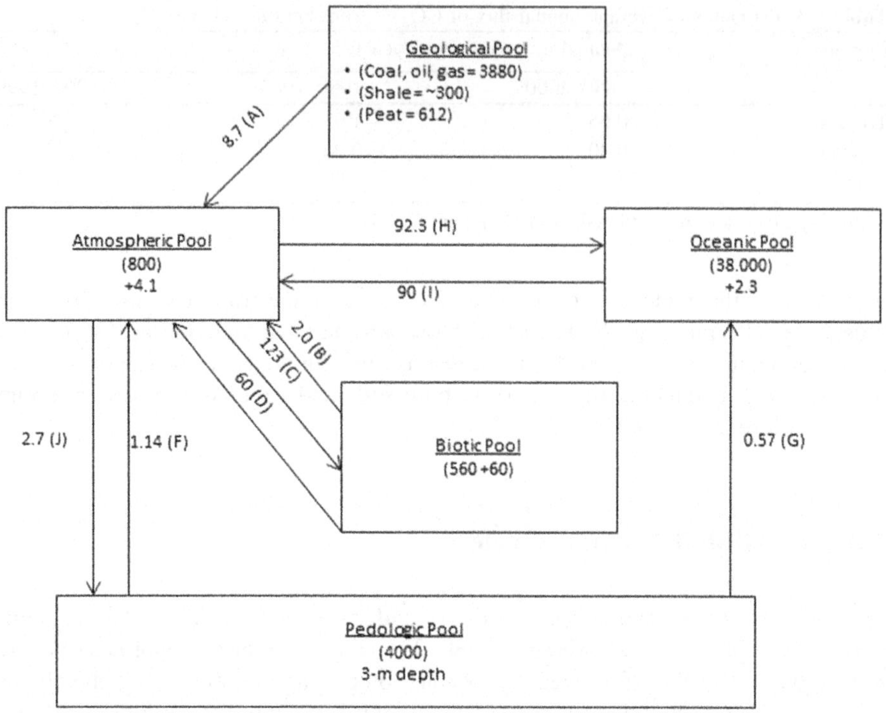

Fig. 1.1 The global carbon cycle. *Source* This author

Table 1.4 Global carbon budget on a decadal scale

Parameter	1960s	1970s	1980s	1990s	2000–2008	2008
1. Sources (Pg C/year)						
Fossil fuel + cement	3.1	4.7	5.5	6.4	7.7	8.7
Land use	1.5	1.3	1.5	1.6	1.4	1.2
Total	4.6	6.0	7.0	8.0	9.1	9.9
2. Sinks (Pg C/year)						
Atmosphere	1.8	2.7	3.4	3.1	4.1	3.9
Ocean	1.5	1.7	2.0	2.2	2.3	2.3
Land	1.2	2.6	1.8	2.6	3.0	4.7
Total known sinks	4.5	7.0	7.2	7.9	9.4	10.9
3. Residual sink	0.1	−1.0	−0.2	0.1	−0.3	−1.0
Total ocean + land sinks	2.8	3.3	3.6	4.7	5.0	5.0
4. Natural sinks as % of sources	60.9	55.0	51.4	58.8	54.9	50.5

Source Adapted and recalculated from LeQuéré et al. (2010)

conversions have decreased at the mean annual rate of 6.25 Tg C/year. Combined, the anthropogenic emissions have increased at the mean annual rate of 0.110 Pg C/year between 1960 and 2008 (Table 1.4).

The atmospheric uptake has progressively increased (Pg C/year) by 1.8 in the 1960s, 2.7 in the 1970s, 3.4 in the 1980s, 3.1 in the 1900s, and 4.1 in the 2000s (2000–2008). The oceanic uptake also increased (Pg C/year) by 1.5 in the 1960s, 1.7 in the 1970s, 2.0 in the 1980s, 2.2 in the 1990s, and 2.3 in the 2000s (2000–2008). The uptake by land-based sinks (Pg C/year) also increased between the 1960s and the 2000s by 1.2 in the 1960s, 2.6 in the 1970s, 1.8 in the 1980s, 2.6 in the 1990s, and 3.0 in the 2000s (2000–2008). The natural sinks (land and ocean combined) absorbed anthropogenic emissions (Pg C/year) by 2.8 in the 1960s, 3.3 in the 1970s, 3.6 in the 1980s, 4.7 in the 1990s and 5.0 in the 2000s. Relative uptake by natural sinks (land plus oceans) as a percentage of the total anthropogenic emission was 61 in the 1960s, 55 in the 1970s, 51 in the 1980s, 59 in the 1990s, and 55 in the 2000s. Some have argued that the C absorption by natural sinks has declined between the 1960s and the 2000s, probably because of the acidification of oceans and degradation and desertification of lands and soils.

Phytosequestration of land plants, and transfer of some *net primary productivity* (NPP) into humus, is a viable option for recarbonizing the biosphere. The annual *gross primary productivity* (GPP) is estimated at 123 Pg (arrow C, Fig. 1.1), of which 60 Pg is returned to the atmosphere through plant respiration (arrow D, Fig. 1.1). Of the remaining 63 Pg, called NPP, around 53 Pg (arrow E, Fig. 1.1) is transferred to roots and allocated to plant metabolism, and the remaining 10 Pg is called the *net ecosystem productivity* (NEP) (Jansson et al. 2010). Most of the NEP is lost to the atmosphere by land use change (arrow B, Fig. 1.1), biotic stresses, fires, and erosion (arrow F, Fig. 1.1). The remainder (0.3–0.5 Pg/year) is called the *net biome productivity* (NBP) (Jansson et al. 2010). The NBP can be enhanced to about 10 Pg C/year through land use and prudent management, and has the potential to persist in the terrestrial biosphere for from centuries to millennia depending on the specific land use in soils and biomass. Thus, C sequestration in the terrestrial biosphere has the potential to offset anthropogenic emissions and mitigate ACC. Here in lies the basic principle of managing the terrestrial C pool to mitigate climate change and also improve the environment.

1.4 The Terrestrial Carbon Pool

The total land area under all biomes, including deserts and ice cover, is 14.3 Bha (Bha = 10^9 ha) (Table 1.5). Of this, 4.85 Bha (33.9 %) is under forest, 2.4 Bha (16.7 %) under savanna, 2.65 Bha (18.5 %) under deserts, 1.88 Bha (13.1 %) under permafrost, 0.8 Bha (5.6 %) under tundra, 0.35 Bha (2.4 %) under peatlands, and 1.4 Bha (9.8 %) under cropland (Table 1.5). The total C pool in vegetation is ~560 Pg, and an additional 60 Pg is contained in the detritus material (see box marked Biotic Pool in Fig. 1.1). Of this, 447 Pg (78.8 %) is contained in forests, 88 Pg (15.5 %) in savanna/grasslands/steppe, and the remaining 32 Pg (5.7 %) in other biomass. The total C pool in the world's soils is estimated at 4,000 Pg to 3 m depth (Table 1.5, see box marked Pedologic Pool in Fig. 1.1).

Of this, 1,104 Pg (27.6 %) is contained in soils under forest, 517 Pg (12.9 %) under savanna, 1,024 Pg (28.6 %) under permafrost, 450 Pg (11.2 %) under peatlands, 332 Pg (8.3 %) under deserts, 144 Pg (3.6 %) under tundra, and 248 Pg (6.2 %) under cropland.

The data in Table 1.5 and the analyses presented above indicate the following:

1. Vulnerability of the pedologic pools to climate change: the projected ACC and the attendant warming may thaw some areas under permafrost and tundra, and also accentuate decomposition of peat lands (through drainage and land use conversion). Thus, 1,618 Pg (40 %) of the pedologic pool is vulnerable to decomposition and emission to the atmosphere with the projected ACC. It is important to identify technological options and policy interventions to minimize the risks of positive feedback from these pools, which comprise 40 % of the total pedologic pools.
2. The last column in Table 1.5 shows the ratio of C density (Mg C/ha) for soil: vegetation. The ratio is 59 in croplands, 36 in tundra, 30 in peatlands, 11 in temperate grasslands, 5 in tropical grasslands and 2.5 in tropical forests. This high ratio implies the high risks of degradation of soils of these ecosystems (by thaw, erosion, fire, deforestation, conversion to other land uses, or drainage) to natural and anthropogenic perturbations. Therefore, the soils of these ecosystems must be managed with extreme caution. An understanding of soil properties and processes is extremely important to the sustainable management of soils of these ecologically-sensitive biomes.
3. Major soils of the world are listed in Table 1.6. In terms of the land area, principal soil Orders include Entisols (16.2 %), Aridisols (12.0 %), Inceptisols (9.8 %), Alfisols (9.6 %), Gelisols (8.6 %), Ultisols (8.4 %), Oxisols (7.5 %), Mollisols (6.9 %), Spodosols (2.6 %), Vertisols (2.4 %), Histosols (1.2 %), and Andisols (0.7 %). Rocky land (10 %) and shifting sands (4.1 %) also cover large areas, but have no or little vegetation cover.

In the context of the SOC pool, the fraction most vulnerable to decomposition or erosion by land misuse and soil mismanagement is that in Gelisols, containing 316 Pg (21 %) of the pedological pools to 1 m depth (Table 1.6). The soils supporting tropical rainforest or Oxisols contain 126 Pg (8.3 %). Molliosols and Histosols together contain 300 Pg or ~20 % of the pedological C pool to 1 m depth. Thus, a total of ~50 % of the terrestrial C pool is vulnerable to decomposition, and may also create positive feedback to ACC.

The strategy of managing C in the terrestrial biosphere is therefore, the following:

1. preserve the existing C pool by minimizing the risks of decomposition and erosion, and
2. enhance the biotic and pedologic pools by carbon sequestration in ecosystems which have been depleted through degradation (erosion, deforestation) and desertification.

Table 1.5 Global distribution of soil and biotic pools in different biomes

Vegetation	Ecoregion	Area (10^6 ha)	Soil carbon (3 m)		Vegetation carbon		Soil C: Veg. C density
			Pool (Pg)	Density (Mg/ha)	Pool (Pg)	Density (Mg/ha)	
I. Forest							
	(1) Tropical	2,450	692	282	276	112	2.5
	(2) Temperate	1,200	262	218	99	82	2.6
	(3) Boreal	1,200	150	91	72	60	1.5
	Subtotal	4,850	1,104	–	447	–	–
II. Savanna/grassland/steppe							
	(1) Tropical	1,500	345	230	72	48	4.8
	(2) Temperate	900	172	191	16	18	10.6
	Subtotal	2,400	517	–	88	–	–
III. Deserts		2,650	332	125	9	3.4	36.8
IV. Peatlands		350	450	1,285	15	43	29.9
V. Permafrost		1,878	1,024	545	–	0	–
VI. Tundra		800	144	180	4	5	36
VII. Cropland		1,400	248	177	4	3	59
Grand total		14,328	4,004	–	567	–	–

Source Adapted from Eglin et al. (2010)

Table 1.6 Estimates of carbon pool in world soils (1 m depth)

Soil order	Area		Soil C pool (Pg)		Total (Pg)
	(10^6 ha)	%	Soil organic carbon	Soil inorganic carbon	
Alfisols	1,262	9.6	158	43	201
Andisols	91	0.7	20	0	20
Aridisols	1,570	12.0	59	456	515
Entisols	2,114	16.2	90	263	353
Gelisols	1,126	8.6	316	7	323
Histosols	153	1.2	179	0	179
Inceptisols	1,286	9.8	190	34	224
Mollisols	901	6.9	121	116	237
Oxisols	981	7.5	126	0	126
Spodosols	335	2.6	64	0	64
Ultisols	1,105	8.4	137	0	137
Vertisols	316	2.4	42	21	63
Rocky land	1,308	10.0	22	0	22
Shifting sand	532	4.1	2	5	7
Total	13,080	100	1,526	940	2,466

Source Adapted from Eswaran et al. (2000)

1.5 Carbon Storage Versus Sequestration

There are two terms commonly used in expressing processes and techniques used in transferring atmospheric CO_2 into other pools with a long MRT. One is *CO_2 Capture and Storage* (CCS) by geoengineering techniques. There are three types of CCS technique: geologic, oceanic, and mineral carbonation (Fig. 1.2). The potential and constraints of these techniques are described by Broecker (2008), Oelkers and Cole (2008), Adams and Caldeina (2008), Benson and Cole (2008), and Schneider (2008) among others.

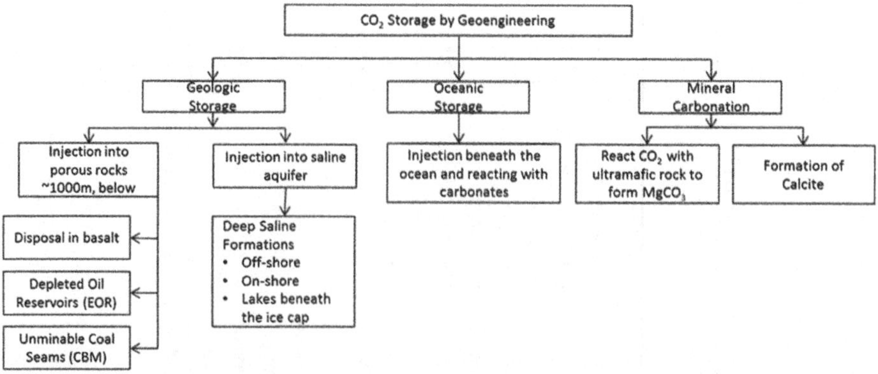

Fig. 1.2 Storage of CO_2 by several geoengineering and geochemical techniques. *Source* This author

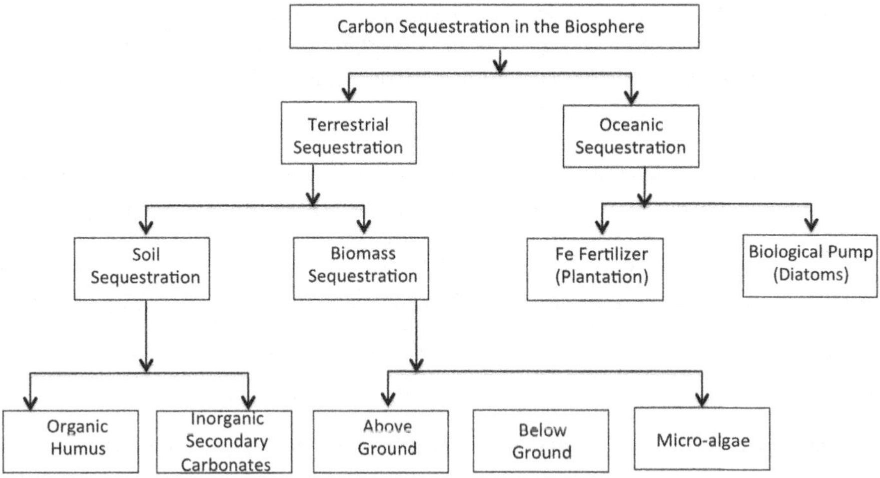

Fig. 1.3 Types of CO_2 sequestration in the biosphere to mitigate climate change. *Source* This author

Table 1.7 Pros and cons of biospheric processes versus engineering techniques of carbon sequestration. *Source* This author

Biotic sequestration	Geoengineering techniques of CCS
1. Pros	**1. Pros**
Natural process	High rate, and large total sink capacity
Cost effective and economical	Rapid process
Numerous co-benefits	Amenable to C trading
Improvement in environment	
No health hazard	
No legal issues	
Amenable to trading C credits	
2. Cons	**2. Cons**
Low sink capacity	Expensive
Dependent on land use	High risks and health hazard
Risks of leakage through decomposition	Few co-benefits (*enhanced oil recovery* (EOR), *coal-bed methane* (CBM))
	Legal issue regarding *measurement, monitoring and verification* (MMV)

In contrast to the geoengineering techniques of CCS, CO_2 sequestration involves the natural processes of photosynthesis and conversion of photosynthate into stable materials (i.e., wood, bio char, humus, recalcitrant organic compounds) so that its MRT within the biosphere is drastically increased from decades to centuries to millennia. Sequestration of CO_2 in the biosphere has two related but distinct components: terrestrial sequestration and oceanic sequestration (Fig. 1.3).

The goal is to retain in the biosphere a large fraction of the NBP and NEP. Details of the CO_2 sequestration techniques in the biosphere are described by Jansson et al. (2010), Jackson and Baker (2010), Lal (2010a), Sayre (2010), Read (2007), and Ogle et al. (2005) among others.

Pros and cons of CCS and biosequestration are outlined in Table 1.7. The natural process of biosequestration is cost-effective, and has numerous ancillary benefits in terms of several ecosystem services. However, CCS techniques are expensive and have high risks, though they are characterized by a large sink capacity. There is no one solution to addressing the complex issue of mitigating ACC by anthropogenic emissions. Niches for each technology must be identified and assessed under site-specific conditions.

1.6 Processes and Techniques of Carbon Sequestration in Soils

The strategy of C sequestration in soil is to increase C gains and reduce C losses. Gains of C in soil are due to addition of biomass from crop residues, animal waste, detritus material from timber and food industry, municipal waste, deposition etc. The loss of C in soil occurs through mineralization or decomposition, erosion, and leaching. The objective is to create a positive C budget in soil, especially by reducing losses through erosion and decomposition. Important techniques of C management on croplands are conservation agriculture or no-till farming, mulching, *integrated nutrient management* (INM) through nutrient cycling and use of bio and synthetic fertilizers, use of complex crop rotations including agroforestry, and application of amendments such as biochar and nano-enhanced materials (zeolites). The importance of soil and water conservation cannot be overemphasized (Lal 2004).

The goal is to produce more from less through sustainable intensification (Lal 2011). The latter implies less but efficient use of energy-based input, especially those with high *hidden C costs* (HCC) such as fertilizers, pesticides, and use of machinery. Conservation and an efficient use of water are also important, especially through replacement of wasteful flood irrigation by micro-irrigation techniques. There are also techniques of management of pasture lands and forest lands (Lal 2010a). Sustainable intensification of a managed ecosystem is needed to meet all the basic necessities of a growing population, regardless of climate change. These basic necessities of the world's population, seven billion in 2011 and projected to reach 9.2 billion by 2050, include an adequate supply of food, feed, fibre, and (bio-)fuel (4 Fs). Increasing the SOC pool to above the threshold level (1.5–2.0 %) in the root zone is essential to improving agronomic production and advancing global food security.

Principal processes of C sequestration in the pedosphere are outlined in Fig. 1.4. There are four basic techniques, two each for the sequestration of SOC

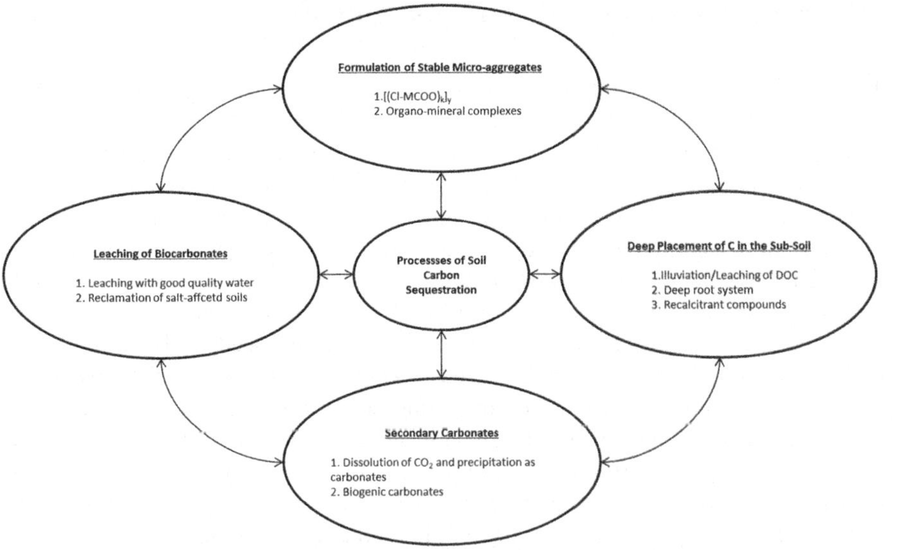

Fig. 1.4 Pedospheric processes leading to sequestration and stabilization of carbon in soil. *Source* This author

and SIC. Enhancing and increasing the stability of structural aggregates (especially micro-aggregates) is a principal mechanism of stabilizing the SOC pool and increasing its MRT. Structural aggregates are created and stabilized through the formation of organo-mineral complexes, especially those involving complexation of organic/humic substances with clay colloids and polyvalent cations (Ca^{2+}, Mg^{2+}, Fe^{3+}, Al^{3+}).

The other mechanism is deep placement of C in the subsoil horizon, away from the climatic elements and the zone of drastic anthropogenic perturbations. In addition to illuviation and leaching, growing perennials/annuals with deep root systems and those containing recalcitrant compounds (e.g. phenols, suberin) can increase the MRT of the C added to the soil solum. Transfer of *dissolved organic C* (DOC) into subsoil and its eventual precipitation in the aquifer or other aquatic ecosystems is also an important mechanism which requires additional research.

Formation of secondary/pedologic carbonates and leaching of bicarbonates are two important mechanisms of the sequestration of SIC (Lal 2004). In general, the rate of C sequestration ranges from 10 to 1,000 kg/ha/year of SOC and from 1 to 10 kg/ha/year for SIC. Management-induced changes in the C pool are sensitive to climate with the following order from largest to smallest changes: tropical moist > tropical dry > temperate moist > temperate dry (Ogle et al. 2005). Thus, management-induced sequestration of SOC is strongly influenced by the present and future climate.

1.7 Sequestration of SOC and Soil Quality

There exists a critical level of SOC below which soil quality is jeopardized (Aune and Lal 1998), it is prone to degradation, and it is less or not at all responsive to inputs. In view of the ever-increasing demands of the burgeoning population, there is a strong need to enhance the SOC pool in the soils of agro-ecosystems. The need is especially urgent for soils of the tropics and subtropics and those managed by resource-poor and small-size landholders who predominately carve out their meagre livings through the widespread use of extractive farming practices. Those soils which have been subject to land misuse and soil mismanagement for a long time, decades to centuries or even millennia, are in dire need of the restoration of their SOC pool. There is an urgent need to enhance agronomic productivity, and this has the additional benefits of offsetting some anthropogenic emissions and improving the environment. For these depleted and degraded soils, each additional 1 Mg C/ha of SOC improves the agronomic yields of crops from 20 to 300 kg/ha, depending on crop, soil type, climate, and management systems (Lal 2006a, 2010b, c). Food production in developing countries can be enhanced by an additional 30 to 50 millions Mg/year by increasing the SOC pool in the root zone of depleted and degraded soils by 1 Mg C/ha (Lal 2006b).

The increase in agronomic productivity by increasing the SOC pool occurs through improvement in soil physical, chemical, and biological quality (Fig. 1.5,

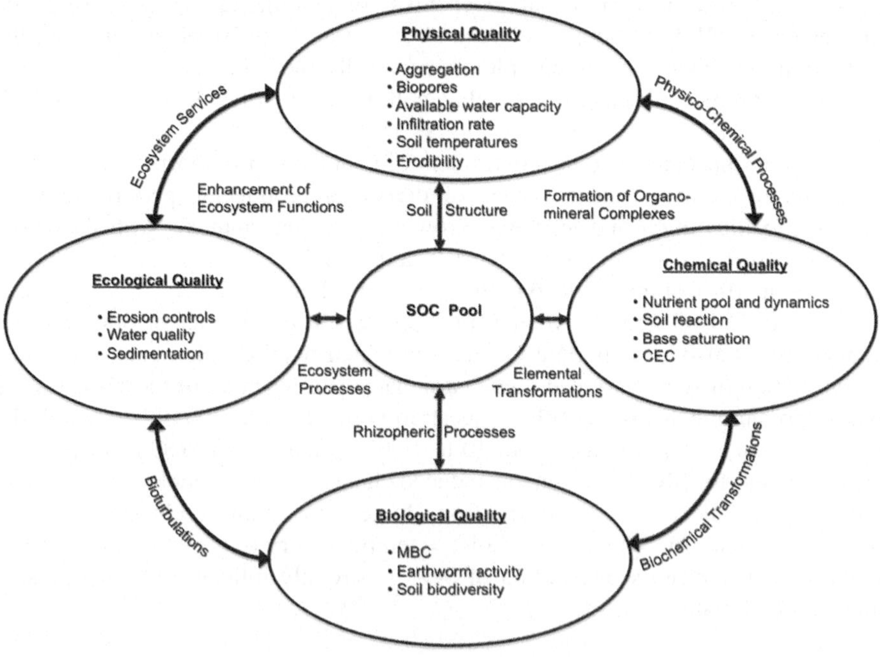

Fig. 1.5 Impacts of improving SOC pool in improvements in soil quality. *Source* This author

Table 1.8 Benefits of restoring organic carbon pool in depleted/degraded soils

Benefits
1. Improving soil structure and tilth
2. Enhancing activity and species diversity of soil fauna
3. Reducing risks of soil erosion and compaction etc.
4. Decreasing non-point source pollution
5. Improving water capacity available to plants
6. Strengthening nutrient cycling
7. Increasing *cation exchange capacity* (CEC) and nutrient retention capacity
8. Increasing agronomic productivity and food security
9. Offsetting anthropogenic emissions
10. Enhancing efficiency of use of agronomic inputs

Source This author

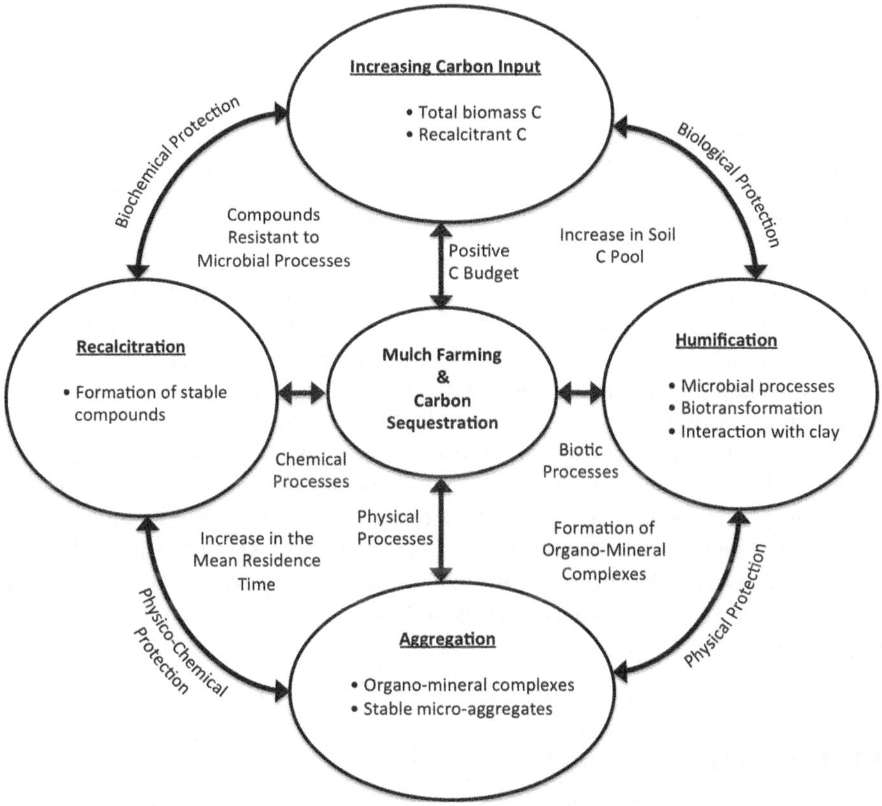

Fig. 1.6 Positive effects of mulch farming techniques on pedologic processes. *Source* This author

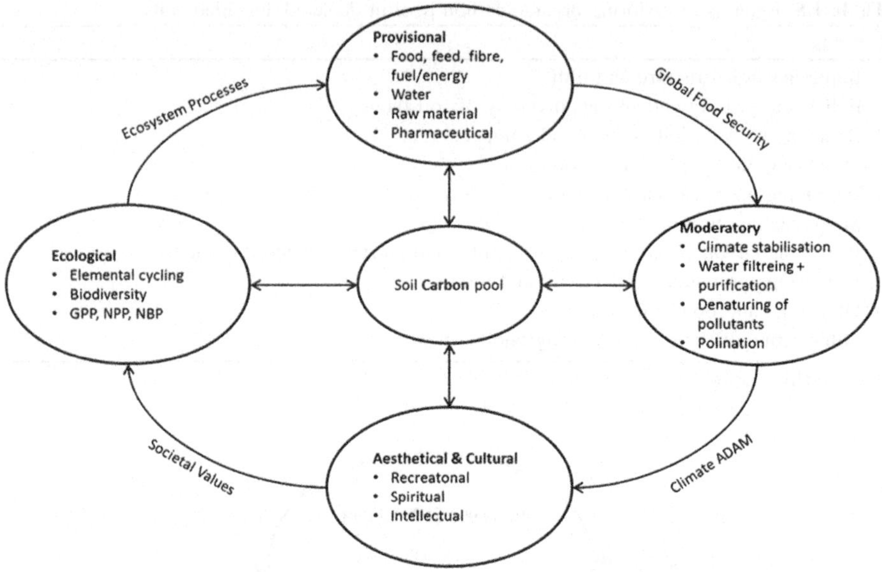

Fig. 1.7 Soil carbon pool and the ecosystem services (*ADAM* = adaptation and mitigation, *GPP* = gross primary productivity, *NPP* = net primary productivity, *NBP* = net biome productivity). *Source* This author

Table 1.8). Among numerous positive effects, improvement in soil structure and the *available water capacity* (AWC) are specifically relevant to physically degraded soils, increase in CEC and nutrient reserves for chemically degraded soils, and increase in biodiversity and *microbial biomass carbon* (MBC) and bioturbation for biologically degraded soils (Fig. 1.5). In addition to an increase in soil quality, long-term adoption of conservation-effective measures based on mulch farming also leads to improvement in biospheric processes (Fig. 1.6).

These processes enhance SOC sequestration, increase MRT, and lead to numerous benefits which improve soil quality (Table 1.8). Both the quality and quantity of SOC pool are drivers of numerous ecosystem services essential to the well-being of the Carbon Civilization and to other vital ecosystem functions (Fig. 1.7).

1.8 Conclusions

Sequestration of C in terrestrial ecosystems, soils, and vegetation is a win–win strategy. In addition to offsetting anthropogenic emissions from fossil fuel combustion and land use conversion, it also generates numerous ecosystem services. Important among these are an improvement in soil quality with an attendant increase in agronomic productivity and use efficiency of inputs, enhancement of the quantity and quality of water resources, and an increase in both above-ground and below-ground biodiversity. The strategy of bio-sequestration differs from that of CCS in being more cost-effective and economical, creating/strengthening numerous ecosystem services, and having low environmental and health-related risks. Improvement in soil quality by increasing the SOC pool is related to beneficial effects on soil physical, chemical, biological, and ecological quality. Common strategies of SOC sequestration are those which create a positive soil C budget by decreasing losses (erosion, mineralization, and leaching), and increasing gains (biomass C). The mean residence time of SOC can be increased by protection against microbial processes through physical, chemical, and biological mechanisms. There is no one silver bullet or panacea, and the choice of appropriate strategies depends on site-specific conditions involving the political parameters governing the issues pertaining to the human dimensions. Regardless of the debate on climate change, recarbonization of the biosphere is essential to the survival of the "Carbon Civilization".

References

Adams, E.E.; Caldeira, K., 2008: "Ocean Storage of CO_2", in: *Elements,* 4: 319–334.

Aune, J.; Lal, R., 1998: "Agricultural Productivity in the Tropics, and Critical Limits of Properties of Oxisols, Ultisols, and Alfisols", in: *Tropical Agriculture,* 74 (Trinidad): 96–103.

Benson, S.M.; Cole, D.R., 2008: "CO_2 Sequestration in Deep Sedimentary Formations", in: *Elements,* 4: 325–331.

Broecker, W.S., 2007: "CO_2 Arithmetic", in: *Science,* 315: 1371.

Broecker, W.S., 2008: "CO_2 Capture and Storage: Possibilities and Perspectives", in: *Elements,* 4: 296–297.

Crutzen, P.J., 2002: "The "Anthropocene"", in: *J. Phys. IV,* 12: 1–5, Doi: 10.1051jp4:20020447.

Eglin, T.P.; Ciais, S.L.; Pias, P., et al., 2010: "Historical and Future Perspectives of Global Soil Carbon Response to Climate and Land Use Changes", in: *Tellus,* 62B: 700–718.

Ellis, C.; Goldewijk, K.K.; Siebert, S., et al., 2010: "Anthropogenic Transformation of the Biomes: 1700 to 2000", in: *Global Ecology and Biogeography,* 19: 589–606.

Eswaran, H.; Reich, P.F.; Kimble, J.M., et al., 2000: "Global Soil Carbon Stocks", in: Lal, R.; Kimble, J.M.; Eswaran, H.; Stewart, B.A. (Eds.): *Global Change and Pedologic Carbonates* (Boca Raton: Lewis Publishers): 15–25.

Foley, J.A.; De Fries, R.; Asner, G.P., et al., 2005: "Global Consequences of Land Use", in: *Science,* 309: 570–573.

Goldewijk, K.K.; Bensen, A.; Van Drecht, G.; de Vas, M., 2011: "The HYDE 3.1 Spatially Explicit Data Base of Human-Induced Global Land Use Change Over the Past 12,000 years", in: *Global Ecology. Biogeography,* 20: 73–86.

Holdren, J.P., 2008: "Meeting the Climate Change Challenge", in: *J.P. Chaffe Memorial Lecture on Science and Environment* (Ronald Regan Blvd, Washington D.C): 17 January, 2008.

Houghton, R.A., 2003: "Revised Estimates of the Annual Net Flux of Carbon to the Atmosphere from Changes in Land Use and Land Management 1850–2000", in: Tellus, 55B: 378–390.

Houghton, R.A., 2007: "Balancing the Global Carbon Budget", in: *Annual Review of Earth and Planetary Sciences,* 35: 313–347.

IPCC, 2000: *Land Use, Land Use Change and Forestry,* Special Report of IPCC (U.K.: Cambridge University Press).

Jackson, R.B.; Baker J.S., 2010: "Opportunities and Constraints for Forest Climate Mitigation", in: *Bioscience*, 60: 698–707.

Jansson, C.; Wullschleger, S.D.; Kalluri, U.C.; Tuskan, G.A., 2010: "Phyto Sequestration: Carbon Biosequestration by Plants and the Prospects of genetic engineering".

Lal, R., 2004: "Soil Carbon Sequestration Impacts on Global Climate Change and Food Security", in: *Science*, 304: 1623–1627.

Lal, R., 2006a: "Enhancing Crop Yield in the Developing Countries Through Restoration of Soil Organic Carbon Pool in Agricultural Lands", in: *Land Degradation & Development*, 17: 197–209.

Lal, R., 2006b: "Managing Soils for Feed a Global Population of 10 Billion", in: *Journal of the Science of Food and Agriculture*, 86: 2273–2284.

Lal, R., 2010a: "Managing Soils and Ecosystems for Mitigating Anthropogenic Carbon Emissions and Advancing Global Food Security", in: *Bioscience*, 60: 708–721.

Lal, R., 2010b: "Enhancing Eco-efficiency in Agroecosystems Through Soil Carbon Sequestration", in: *Crop Science*, 50: S120–S131.

Lal, R., 2010c: "Beyond Copenhagen: Mitigating Climate Change and Achieving Good Security Through Soil Carbon Sequestration", in: *Food Security*, 2: 169–177.

Lal, R., 2011: "Harnessing Science Knowledge for Combating Desertification, Land Degradation and Drought", Keynote Paper Presented at the 10th Session of COP to UNCCD, Changwan, South Korea, 17–18 October.

LeQuéré, C.; Raupach, M.R.; Canadell, J.G., et al., 2010: "Trends in Source and Sinks of CO_2", in: *Nature Geosciences* (www.nature.com/naturegeoscience).

Marland, G.; Rotty, R.M., 1984: "Carbon dioxide Emissions from Fossil Fuel: A Procedure for Estimation and Results for 1950–1982", in: *Tellus*, 36(B): 232–261.

Marland, G.; Boden T.A.; Andre, R.J., 2007: "Global, Regional, and National CO_2 Emissions", in: *Trends: A compendium of Data on Global Change: CO_2 Information Analysis Center* (Oak Ridge, TN: ORNL).

McNeill, J.R., 2000: *Something New Under the Sun* (N.Y: W.H. Norton and Co.): 421.

Oelkers, E.H.; Cole, D.R., 2008: "Carbon dioxide Sequestration: A Solution to a Global Problem", in: *Elements*, 4: 305–310.

Ogle, S.M.; Breodt, F.J.; Paustian K., 2005: "Agricultural Management Impacts on Soil Organic Carbon Storage Under Moist and Dry Climatic Conditions of Temperate and Tropical Regions", in: *Biogeochemistry*, 72: 87–121.

ORNL., 2001: *Global CO_2 Emissions from Fossil Fuel Burning, Cement Manufacture, and Gas Flaring* (Tennessee, USA: Oakridge National Lab): 1751–1998.

Ramanathan, V.; Xu, Y., 2010: "The Copenhagen Accord for Limiting Global Warming: Criteria, Constraints, and Available Avenues", PNAS, (pnas.org./cgi/doi/10.1073/pnas.100229317).

Read, P., 2007: "Biosphere Carbon Stock Management: Addressing the Threat of Abrupt Climate Change in the Next Few Decades: An Editorial Essay", in: *Climatic Change*, (doi: 10.1007/s10584-007-9356-y).

Ruddiman, W.F., 2003: "The Anthropogenic Greenhouse Era Began Thousands of Years Ago", in: *Climatic Change*, 61: 261–293.

Ruddiman, W.F., 2006: "On the Holocene CO_2 Rise: Anthropogenic or Natural?", in: *EOS*, 87: 352–353.

Sayre, R., 2010: "Micro-Algae: The Potential for Carbon Capture", in: *Bioscience*, 60: 722–728.

Schneider, S.H., 2008: "Geoengineering: Could we or Should we Make It Work?", in: *Philosophical Transactions of the Royal Society (A)*, 366: 3843–3862.

Sterman, J.D., 2008: "Risks Communication on Climate: Mental Models and Mass Balance", in: *Science*, 322: 532–533.

Trenbath, K.E.; Christy, J.R.; Olson, J.G., 1988: "Global Atmospheric Mass, Surface Pressure, and Water Vapor Variations", in: *Journal of Geophysical Research*, 93D: 10925.

UNFCCC., 2009: *Copenhagen Accord*, (htttp://unfccc.int/resources/docs/2009/cop15/eng/107.pdf).

Chapter 2
Land Degradation and Security Linkages in the Mediterranean Region

Winfried E. H. Blum

Abstract Impacts of land degradation on the technical, social, economic, and cultural environments of the Mediterranean region are discussed under their security aspects, including the risks caused by climate change and the possibilities of mitigation.

Keywords Land degradation · Mediterranean · Security · Climate change · Politics

2.1 Introduction

Land degradation in regions with marked water scarcity, like the Mediterranean region, is seen as a trigger for security problems in the social, economic, and cultural context. For this reason, NATO, together with the OSCE, organized an international conference in Valencia, Spain in late 2009 to discuss the security aspects of the impacts of land degradation, desertification, and water scarcity in the Mediterranean region (Rubio et al. 2009). In the following paper, some of the results related to soil and land degradation will be reported (see also Blum 2009; Kapur et al. 2011).

2.2 Land Degradation

"Land" normally means a physical entity in terms of its topography and spatial nature, including the natural resources such as soils, minerals, water, and biota that the land comprises (UNEP 2001). The six main forms of land use are described by Blum (2005).

W. E. H. Blum (✉)
Institute of Soil Research, University of Natural Resources and Life Sciences, Vienna, Austria
e-mail: winfried.blum@boku.ac.at

S. Kapur and S. Erşahin (eds.), *Soil Security for Ecosystem Management*,
SpringerBriefs in Environment, Security, Development and Peace 8,
DOI: 10.1007/978-3-319-00699-4_2,

Land degradation is mainly caused by two types of unsustainable land use:

- sealing of land through urbanization and industrialization, excluding all further uses of soil and land such as biomass production, filtering, buffering, and transformation, as well as the function of soil as a gene reserve;
- unsustainable use of specific land functions, such as tourism and agriculture, causing erosion, compaction, contamination, and further soil deterioration.

The problem of sealing becomes visible by comparing Europe's natural resources at daytime with Europe's built environment at night, showing that large parts of Europe are sealed by urbanization, industrialization, and transport, from which emissions are released on the adjacent land surfaces (see Figs. 2.1 and 2.2). Figure 2.3 gives a detailed view of the increase in artificial areas in the coastal zones of the Mediterranean region between 1975 and 1990 by percentage, and the projected increase in urban population between 1990 and 2025 (EEA 2001).

These pictures indicate that sealing prevents all other uses of soil and land, especially biomass production, filtering of rain water, buffering, and biological transformation, as well as its function as a gene reserve.

Figure 2.4 shows in more detail the process of sealing of a landscape, taking southern Germany as an example, with towns, villages, and roads of first and second order connecting urban and peri-urban settlements, at a scale which is indicated in the picture.

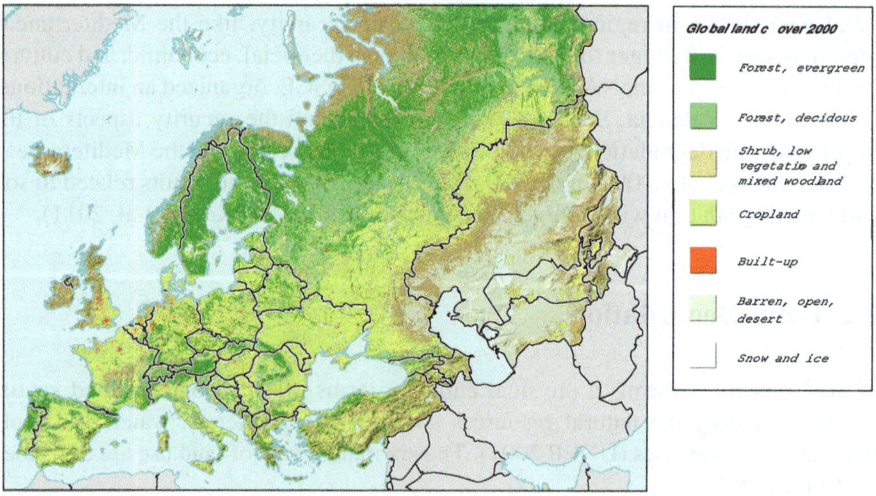

Fig. 2.1 European natural resources in daylight. *Source* http://www.eea.europa.eu/data-and-maps/figures/distribution-of-natural-resources-in-the-pan-european-region-for-selected-issues-2

Fig. 2.2 Europe's built environment at night. *Source* http://www.en.wikipedia.org

Sealing for urbanization, industrialization, transport, and tourism is causing security problems in two ways:

1. Ecological-technical problems are:
   - impedance of rainwater infiltration, causing surface run-off, with the danger of flooding and the loss of rainwater storage in areas where this water would be urgently needed;
   - high evaporation and water losses to the atmosphere from urban surfaces sealed by asphalt, concrete, and other dense materials, such as roofs, streets, and parking lots;
   - increased temperature levels due to storage of radiation energy in constructions;
   - production and accumulation of refuse and emission of dust and gases; and
   - increased and concentrated demand for water in competition with other uses, e.g. agriculture.

   For this reason, extensive tourist areas in the Mediterranean basin must be regarded from the aspect of security linkages.

2. Social, economic, and cultural problems are caused by:
   - disappearance of natural landscapes formerly used for recreation, agriculture, and forestry;

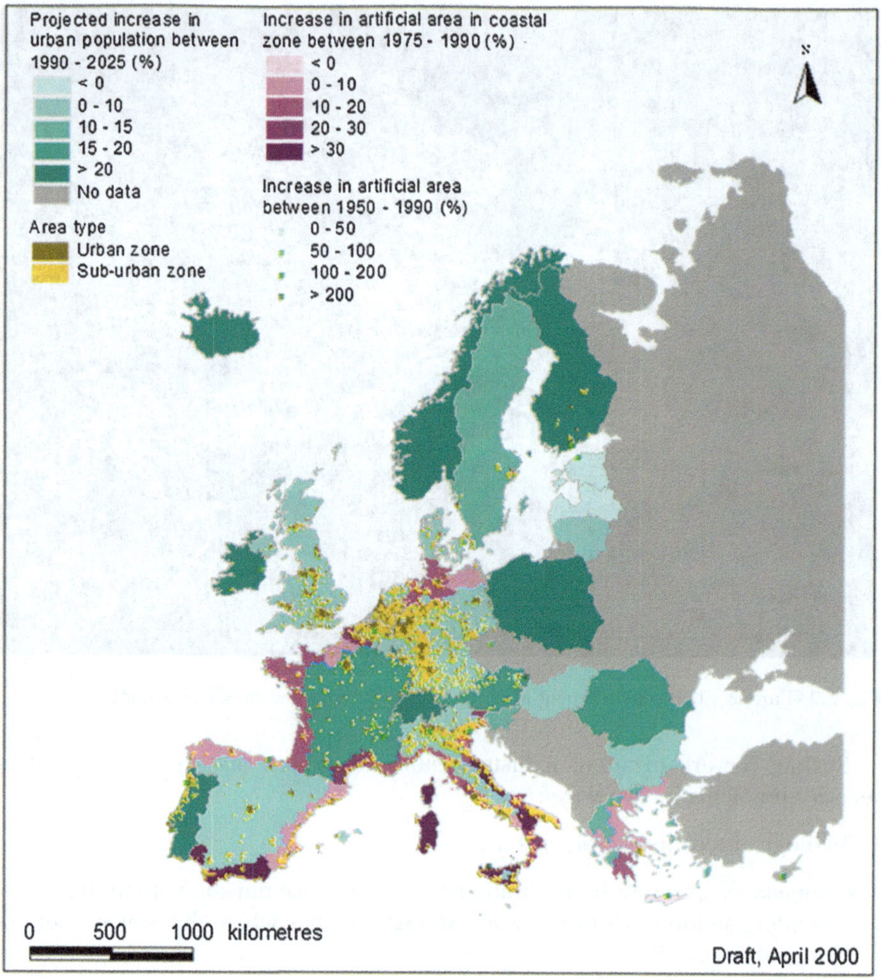

Fig. 2.3 Increase of artificial areas in the Mediterranean coastal zone 1975–1990 in %. *Source* http://www.eea.europa.eu

- loss of livelihood for farmers and pastoralists through the loss of crop- and pastureland; and
- emergence of new social groups, especially in urban and peri-urban agglomerations, with problems of integrating into the existing social and economic environments.

Besides sealing as an exclusive form of competition, urban and peri-urban agglomerations cause important impacts, through physical and chemical loads, on the adjacent agricultural and forest lands, on the atmospheric pathway, on the waterways, and through terrestrial transport (see Fig. 2.5). These processes are still going on and have even been accelerating in the last decade, contaminating land

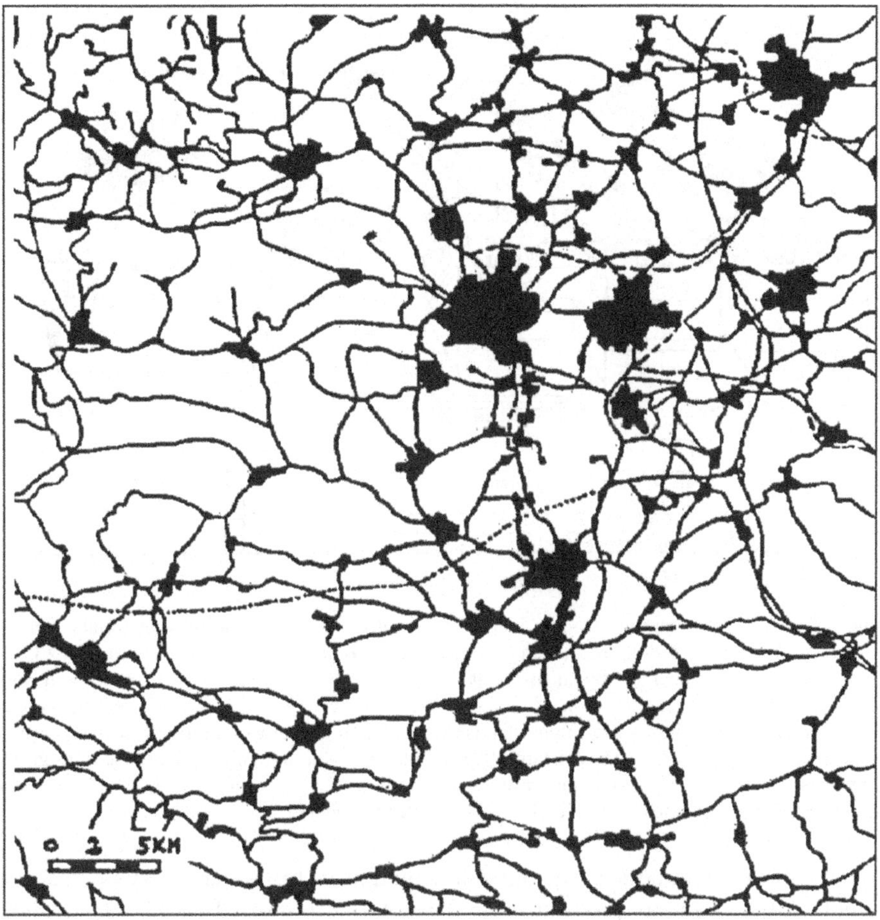

Fig. 2.4 Sealing the landscapes in south-western Germany (note scale of 5 km). *Source* This author

and water surfaces with heavy metals and toxic organics at an intensity never observed before (Blum 1998, 2006).

An extreme form of land degradation is desertification, occurring under arid and semi-arid climatic conditions, characterized by a lack of rainfall during long periods of the year and a deficit in the water balance. These areas are specifically vulnerable to land degradation caused by forest fires, overgrazing, unsustainable agricultural cropping or urbanization, and industrialization (Blum 2006).

Desertification is mostly related to water scarcity caused by a severe water deficit. Moreover, a special problem in areas with water deficit is salinization through irrigation without sufficient drainage. Further problems occur through imbalanced water availability in upstream and downstream areas of water reservoirs. People upstream often have limitations on their land use, being forced to

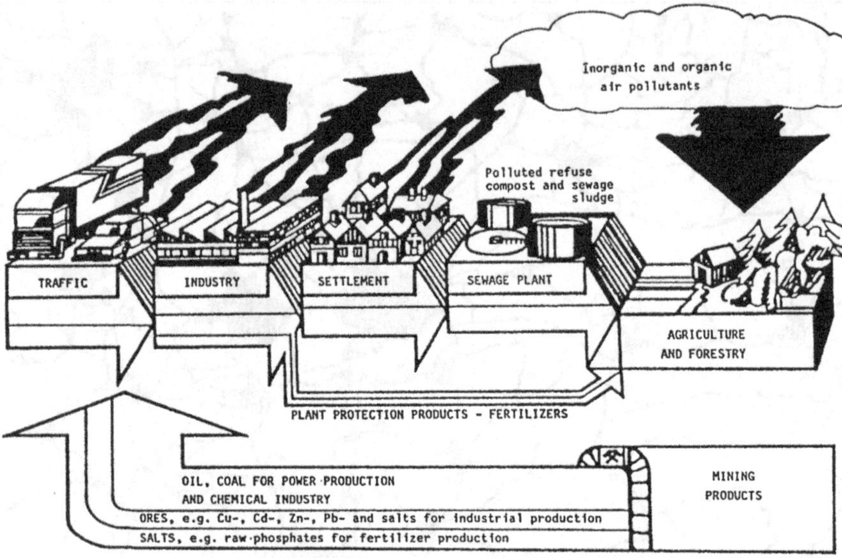

Fig. 2.5 Soil contamination through excessive use of fossil energy and raw materials. Further impacts of land degradation are caused by tourism (Previtali 2011), mining activities (Faz Cano et al. 2011), and especially by unsustainable agriculture. Figure 2.6 shows impacts on soils through compaction, accumulation of contaminants, e.g. pesticides, through the use of manures and fertilizers, through sewage sludge deposition, through soil erosion, through loss of organic matter, and through other causes (Blum 2007a). Unsustainable agricultural land use through the centuries can change and even ruin entire landscapes, as can be seen in many areas north and south of the Mediterranean (cf. Fig. 2.7). *Source* Blum (1988)

avoid soil erosion and subsequent sedimentation of the water reservoir, whereas people downstream are profiting from the water accumulated by the dams. Under such conditions conflicting interests exist, and these have to be solved by politics and decision-making.

Recently, new conflicts have arisen though the competition between the production of food on one side and of biofuels such as ethanol and biodiesel on the other, aggravated through water scarcity since water is needed for both production lines.

Looking into the impacts in order of urgency reveals that land and soil losses through sealing, mining of soil material, and soil erosion by water and wind, as well as intensive pollution of soils by heavy metals, xenobiotics, radioactive compounds, and advanced salinization and acidification and deep-reaching soil compaction are irreversible, in the sense that they cannot be reversed in a time span of about a hundred years or four human generations.

This judgment is important for defining priorities in combating land degradation and related security problems which may increase and accelerate with time, causing large-scale security issues.

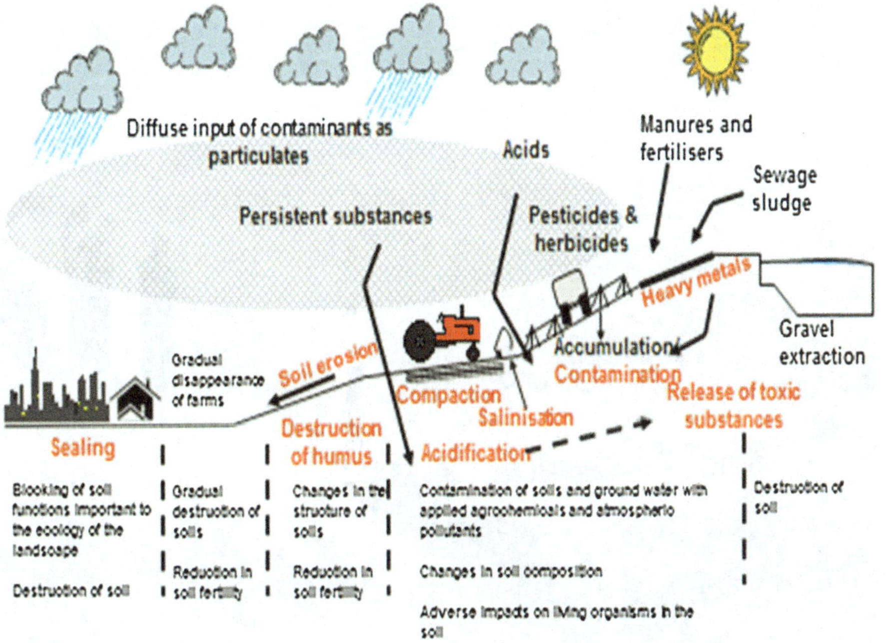

Fig. 2.6 Impact of agricultural and other human activities on soil. *Source* Eu.JRC.IES.Ispra/Italy

2.3 Security Problems Related to Water Scarcity

Water scarcity is typical for countries in the Mediterranean Basin because of its specific climatic conditions. Competition for the use of the scarce water resources already exists in areas with intensive agricultural irrigation, which consumes on the average more than 70 % of all available freshwater resources and leaves only less than 30 % for domestic and industrial purposes, not counting severe water losses through deficient water distribution systems such as pipes, where often up to 25 % of all available water is lost. The share of irrigated arable land in the Mediterranean region can be seen in Fig. 2.8 (FAO 2006).

Under these conditions, tourist activities during the dry season with a high demand for water can come into conflict with the needs of the local population, especially for agriculture and gardening. Moreover, unsustainable water use through agriculture with insufficient irrigation techniques causes loss of water reserves and salinization, and has an additional adverse impact on estuarine and coastal environments.

In some Mediterranean regions, it can increasingly be observed that rivers are no longer supplying groundwater resources. On the contrary, the groundwater resources are providing water for the rivers, which is a strong signal that the water balance is heavily disturbed. These processes are part of the desertification process

Fig. 2.7 Extreme land degradation through long-term agricultural activities in western Algeria: *Source* Photograph by the author

and raise important issues of security, because under the prevalent and possible future climatic conditions, the results may be irreversible and can therefore endanger the basis of life of large parts of the population (Blum et al. 2004).

To summarize, security linkages can arise from two different kinds of impact:

- natural environmental imbalance leading to natural disasters such as extreme meteorological conditions, forest fires, and landslides;
- human-induced environmental impacts such as the depletion of natural resources, especially of soil and water, the loss of biodiversity, the sealing of land by urbanization, industrialization and tourism, and as an overall result of human activities climate change, which is a global result of innumerable locally defined processes.

For their impact as well as for the security problems they cause, two parameters are of paramount importance:

- the dimension of space, meaning, for example, the spatial scale of urbanization, of soil contamination, of landslides, and of forest fires and other events;
- the dimension of time, meaning the pace at which the impact-driven processes occur. For example, it is very important to determine whether sealing of large areas by urbanization occurs within a few years or within decades.

In this context, the impact of climate change might cause severe future risks.

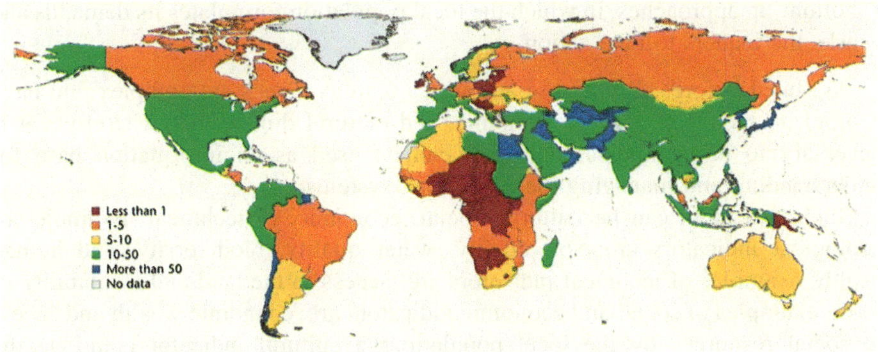

Fig. 2.8 Share of irrigated arable land. *Source* www.unesco.org

2.4 Climate Change and Future Risks

Looking at the *Intergovernmental Panel on Climate Change* (IPCC) scenario up to 2100, it becomes clear that not only CO_2 emissions but also the increase of CO_2 in the atmosphere and the resultant temperature rise are severe threats, especially to the Mediterranean region (see also Steffen et al. 2004). The overall indications are that the mean annual temperature in the Mediterranean Basin will increase, whereas the mean annual precipitation will decrease. A special case study for the Mediterranean coastal area is given by Fujinawa (2011).

If the distribution of irrigated land in the Mediterranean Basin is considered (Fig. 2.8), it becomes clear that in the Mediterranean region, several countries, for example Egypt, which is almost totally dependent on irrigation, will severely suffer from the decrease in water resources (see also Parry et al. 2004; Fischer et al. 2005).

2.5 Solving Security Issues by Bridging Between Science, Politics, and Decision-Making

Impacts on social and economic systems caused by land use and water scarcity and the resulting conflicts have to be controlled by politics and decision-making, because sustainable use of land, e.g. through spatial and/or temporal harmonization of the uses in a given area, to avoid or minimize irreversible impacts, as well as imbalanced distribution of water resources, is not a scientific but a political issue (Blum 2006, 2007a, b). Two decision patterns can be distinguished:

- top-down decisions, in which the top-ranking decision-makers manage the infrastructure;

- bottom-up approaches, in which the local population formulates its demands and asks the leaders to take action.

As stated before, all land use and water scarcity issues are complex and have ecological, technical, social, economic, and cultural dimensions. Therefore, it is necessary to define indicators which can be used as an information base for understanding and managing these complex systems.

Such indicators can be cultural, social, economic, or technical. Examples of ecological indicators are: soil quality, water quality, biodiversity, and human health; examples of technical indicators are: access to the land and availability of tools; examples of social and economic indicators are: economic wealth and access to social resources by the local population; a cultural indicator could be the educational level in a certain region (Blum 2004).

The criteria for these indicators are fourfold:

- they must be policy-relevant and focus on the real demands;
- they must be analytically sound, based on good science, and demonstrate a clear cause-effect relationship;
- they must be easy to interpret and understandable by farmers and stakeholders at the grassroots level as well as by decision-makers and politicians;
- they must be easily measurable and therefore feasible and cost effective in data collection, data processing, and dissemination.

With the help of such indicators it should be possible to mitigate and alleviate future security problems, and so diminish the security linkages caused by unsustainable land use and water scarcity in the Mediterranean region.

References

Blum, W.E.H., 1988: Problems of Soil Conservation. Nature and Environment Series 39, Council of Europe, Strasbourg, France.

Blum, W.E.H.; Barcelo, D; Büsing, J; Ertel, T; Imeson, A; Vegter, J., 2004: Scientific Basis for the Management of European Soil Resources. Research Agenda (Verlag Guthmann-Peterson, Wien 2004): pp 18. (ISBN 3-900782-47-4).

Blum, W.E.H., 1998: "Soil Degradation Caused by Industrialization and Urbanization", in: Blume, H.-P.; Eger, H.; Fleischhauer, E.; Hebel, A.; Reij, C.; Steiner, K.G., (Eds.): *Towards Sustainable Land Use, Advances in Geoecology 31*, Vol. I (Catena Verlag, Reiskirchen): 755–766.

Blum, W.E.H., 2004: "Soil Indicators for Decision Making—Sharing Knowledge Between Science, Stake Holders and Politics", 13th International Soil Conservation Organisation Conference, 4–8 July 2004, Brisbane/Australia. Conference Proceedings (CD), Paper No. 202, pp. 1–5, Brisbane.

Blum, W.E.H., 2005: "Functions of Soil for Society and the Environment", in: *Reviews in Environmental Science and Biotechnology*, 4: 75–79.

Blum, W.E.H., 2006: "Urban and Peri-Urban Environments: Emerging Frontiers in Soil and Water Conservation". E.S.S.C. Newsletter 3: 16–23, Wolverhampton/UK.

Blum, W.E.H., 2007a: "Role of Soils in River Basin Management", in: *RiskBase 1st Thematic Workshop*, May 17–18, 2007, Lisbon, Portugal, pp. 15–18. IIQAB-CSIC, Barcelona, Spain.

Blum, W.E.H., 2007b: "From Vulnerability to Resilience—Search for New Concepts, Explaining the Need for Integrated Risk-Based Management of the Water-Sediment-Soil System at River-Basin Scale". RiskBase-Newsletter 1, p. 2, Oct 2007, TNO, Utrecht/NL, 2007b.

Blum, W.E.H., 2009: "Reviewing Land Use and Security Linkages in the Mediterranean Region", in: Rubio, Safriel, Daussa, Blum, Pedrazzini (Eds.): *Water Scarcity, Land Degradation and Desertification in the Mediterranean Region* (Dordrecht: Springer): 101–117.

EEA, 2001: *Environmental signals 2001. Environmental assessment report No. 8* (European Environment Agency, Copenhagen).

FAO, 2006: Land and Water Digital Media Series 11. Global Agro-ecological Zones, Version 1.0. IIASA Science for Global Insight, 2006.

Faz Cano, A.; Zanuzzi, A.; Martinez-Pagan, P.; Acosta, J.; Carmona, D.; Martinez–Martinez, S.; Muńoz, M., 2011: "Roman Mining Landscapes in the Murcia Region, SE Spain: Risk Assessment of Mine Ponds", in: Kapur, S.; Eswaran, H., Blum, W.E.H., (Eds.): *Sustainable Land Management* (Springer Verlag Berlin, Heidelberg): 293–310.

Fischer, G.; Shah, M.; Tubiello, F.N.; Van Velhuizen H., 2005: "Socio-economic and Climate Change Impacts on Agriculture: An Integrated Assessment, 1990–2080", in: *Philosophical Transactions of the Royal Society B*: 360:2067–2083.

Fujinawa, Katsuyuki, 2011: "Anthroscape of the Mediterranean Coastal Area in the Context of Hydrogeology: Projected Impacts of Climate Change", in: Kapur, S.; Eswaran, H.; Blum, W.E.H., (Eds.): *Sustainable Land Management* (Springer-Verlag Berlin, Heidelberg): 143–162.

Kapur, S.; Eswaran, H.; Blum, W.E.H., (Eds.): *Sustainable Land Management—Learning from the Past for the Future.* (Springer Heidelberg, Dordrecht, London, N.Y.) 2011 (ISBN 978-3-642-14781-4).

Parry, M.L.; Rosenzweig, C.; Iglesias, A.; Livermore, M.; Fischer, G., 2004: "Effects of Climate Change on Global Food Production Under SRES Emissions and Socio-Economic Scenarios", in: *Global Environmental Change*, 14: 53–67.

Previtali, F., 2011: "Mountain Anthroscapes, the Case of the Italian Alps", in: Kapur, S.; Eswaran, H.; Blum, W.E.H., (Eds.): *Sustainable Land Management* (Springer-Verlag Berlin, Heidelberg): 143–162.

Rubio, J.L.; Safriel, U.; Daussa R.; Blum, W.E.H.; Pedrazzini, F. (Eds.), 2009: Water Scarcity, Land Degradation and Desertification in the Mediterranean Region. (Springer, Dordrecht), ISBN 978-90-481-2525-8(PB).

Steffen, W.; Sanderson, A.; Tyson, P.D.; Jäger J.; Matson, P.A.; Moore, III B.; Oldfield, F.; Richardson, K.; Schellnhuber, H.J.; Turner, II B.L.; Wasson, R.J., 2004 : "Global Change and the Earth System—A Planet Under Pressure". (Springer-Verlag Berlin Heidelberg New York), ISSN 1619-2435, ISBN 3-540-40800-2.

UNEP, 2001: *Global Ministerial Environment Forum Policy Issues: State of the Environment.* UNEP's Policy on Land and Soil. UNEP/GC21-IDF-13, January 2001, Nairobi.

[illegible]

Chapter 3
The Role of Soil Information in Land Degradation and Desertification Mapping: A Review

Claudio Zucca, Riccardo Biancalani, Selim Kapur, Erhan Akça, Pandi Zdruli, Luca Montanarella and Freddy Nachtergaele

Abstract Mapping *land degradation and desertification* (LDD) has generally been considered as a complex task, and past efforts have produced contrasting results. Until recently, this exercise has often been seen as a soil scientist's task by the international community. However, the actual role and "weight" of soil information in LDD mapping at different spatial scales has been influenced and constrained by the changing conceptual frameworks and data availability. This chapter reviews these aspects and discusses the most recent developments. Starting from the evolving definitions of land degradation and desertification, it describes the use made of soil information by past global mapping initiatives. It presents the related past and new conceptual frameworks, and describes the approaches adopted by the most relevant ongoing international initiatives such as LADA and WAD. Finally, it highlights the existing constraints and limitations and provides recommendations on gaps and needs in terms of soil-related knowledge and data.

C. Zucca (✉)
Dipartimento Di Agraria, University of Sassari, V.le Italia 39, 07100 Sassari, Italy
e-mail: clzucca@uniss.it

R. Biancalani · F. Nachtergaele
FAO, Viale delle Terme di Caracalla, 00153 Rome, Italy
e-mail: Riccardo.Biancalani@fao.org

S. Kapur
Department of Soil Science and Archaeometry, University of Çukurova, 01330 Adana, Turkey

E. Akça
Technical Programs, Adiyaman University, Adiyaman, Turkey

P. Zdruli
International Centre for Advanced Mediterranean Agronomic Studies, Mediterranean Agronomic Institute of Bari (CIHEAM IAMB), Bari, Italy

L. Montanarella
European Commission, DG JRC, Ispra, Italy

S. Kapur and S. Erşahin (eds.), *Soil Security for Ecosystem Management*,
SpringerBriefs in Environment, Security, Development and Peace 8,
DOI: 10.1007/978-3-319-00699-4_3, © The Author(s) 2014

Keywords LDD · Desertification · Land degradation mapping · Soil status indicators · Soil data

3.1 The Evolving Concept and Definition of Land Degradation and Desertification

Since its coming into force, the *United Nations Convention to Combat Desertification* (UNCCD 1994) has stimulated and framed the international debate on *land degradation and desertification* (LDD). Desertification is defined by the Convention as "land degradation in drylands": in other words, desertification is a climatic sub-type of land degradation. However, the long-lasting debate that led to the UNCCD definition is still developing, and in need of a more science-based and rigorous definition. The related conceptual evolution has been discussed by several authors (Thomas and Middleton 1994; Eswaran et al. 2001; Le Houérou 2002; Herrmann and Hutchinson 2005; Safriel 2007; Reynolds et al. 2007).

The most recent tendency is towards defining desertification as "an end state of the process of land degradation, expressed by a persistent reduction or loss of biologic and economic productivity of lands" (DSD 2009). Its causes would be linked to both human and natural factors; these often act synergistically.

The underlying basic concept of "land degradation" has evolved over time too. Over the last thirty years land degradation concepts have moved from an initial emphasis on the productive capability of the soils to the all-encompassing concept of ecosystem capacity to provide goods and services to society. Some of the most recent definitions are listed below.

- The reduction or loss of the biological or economic productivity and complexity of the land resulting from land uses or from a process or combination of processes, including processes arising from human activities and habitation patterns (UNCCD 1994).
- The reduction in the capacity of the land to provide ecosystem goods, functions, and services that support society and development (MEA 2005).
- The reduction in the capacity of the land to provide ecosystem goods and services over a period of time for its beneficiaries (Nachtergaele et al. 2011).

According to the UNCCD, "land" means "the terrestrial bio-productive system that comprises soil, vegetation, other biota, and the ecological and hydrological processes that operate within the system".

Current ratification of the UNCCD reflects such a broader definition of desertification with the vast majority of the countries of the world declaring themselves as affected by desertification processes (Table 3.1). Among the 165

Table 3.1 Parties to the UNCCD and their status as affected or non-affected by desertification at the end of 2009. Roman numerals identify the five UNCCD Regional Annexes. *Source* The authors

Annexe	Affected	Non-Affected	Total no. of countries
1: Africa	53	0	53
2: Asia	51	5	56
3: LAC	33	0	33
4 and 5: Northern Mediterranean and/or Central and Eastern Europe	25	4	29
Non-annexe-specific OECD countries	3	19	22
Total	165	27	193

affected country Parties, 154 prepared a report as affected countries, and 103 have prepared their *National Action Programme* (NAP).[1]

Of the 193 UNCCD countries, 162 are developing countries (of which, 93 dryland-affected and 69 non-dryland-affected countries). The ratio of non-dryland to dryland developing countries rose from 0.33 in 1995 to 0.74 in 2004. Safriel (2007) discusses these figures and links this tendency with the increased funding opportunities that derive from institutions like the *Global Environment Facility* (GEF) or the *World Bank* (WB) that have included in their portfolios consistent amounts of funding for combating land degradation. Therefore, the tendency is to look at the UNCCD as a convention that deals with land degradation not just in the drylands but in all lands that could be at risk of degradation or are already degraded. This would represent a major shift for the convention and would give it a more global mandate.

Safriel (2009) discusses the links between desertification and drought and concludes that they are not straightforward and that the former does not necessarily precede the latter, illustrating this with examples from the Sahel droughts of 1968–1974 and 1983–1985. Again, the adoption of the DLDD (*desertification, land degradation and drought*) terminology shows that the UNCCD is struggling to adjust its actions to reflect its wider membership.

Obviously soil degradation, soil quality, and soil health play a crucial role in identifying areas affected by land degradation and desertification, and have been given varying definitions. The *World Reference Base for Soil Resources* (WRB) (FAO/ISRIC/ISSS 2006) describes the soil as:

> …any material within 2 m from the Earth's surface that is in contact with the atmosphere, with the exclusion of living organisms, areas with continuous ice not covered by other material, and water bodies deeper than 2 m.

Soil is the interface between earth, air and water and hosts most of the biosphere. It is essentially a non-renewable resource in that the degradation rates can be rapid whereas the formation and regeneration processes are extremely slow.

[1] See at: <http://www.unccd.int/en/about-the-convention/Action-programmes/Pages/default.aspx>.

Soil degradation is described by *physical, chemical, and biological degradation* processes acting upon the soil and impacting human well-being and livelihoods (Kaiser 2004). Each of these categories includes a set of degradation processes as described below:

- physical degradation: decline in soil structure, crusting, compaction, sealing, and erosion;
- chemical degradation: acidification, leaching, nutrient depletion, salinization, and alkalinization;
- biological degradation: depletion of soil organic matter, reduction of soil biodiversity, decline in activity and species diversity of soil fauna and flora.

A combination of some of these threats can ultimately propel arid or sub-arid climatic conditions towards desertification (European Commission 2006).

Given the conceptual differences between "land" and "soil", it is important to point out the distinction between "land quality" and "soil quality". Eswaran et al. (2003) state that the concept of soil quality is acceptable for plot- and farm-scale assessments, while land quality is appropriate for more general evaluations referring to national, regional, continental, or global scales. Land degradation often initiates when there is a mismatch between land quality and land use.

The same authors introduce the concept of *inherent land quality* (ILQ) and describe it as "the ability of land to perform its functions under natural conditions influenced only by the intrinsic properties of the ecosystem and not significantly modified by land management" (Eswaran et al. 2003). In addition to this concept they mention that of *managed land quality* (MLQ), which refers to the ability of land to function under managed conditions. The latter is very important since very few places on Earth have remained untouched by humans. The comparison between ILQ and MLQ indicates whether the natural equilibrium has been disturbed towards enhancement or degradation.

Another interesting concept that needs attention is *land resilience,* described as "the ability of land to restore to an acceptable level of performance subsequent to degradation" (Eswaran 1994). Resilience depends on soil-intrinsic fertility conditions as well as on the socio-economic and political situation where the land unit under consideration lies.

3.2 Recent Conceptual Developments

3.2.1 Ecosystem Goods and Services (G&S)

The ecosystem services framework is increasingly thought to provide a basis for assessing and valuing the impacts of land change and degradation, as well as the effects of the actions aimed at reversing it (DSD 2009). The major LDD processes, including water and wind erosion, soil salinization, loss of vegetation cover and diversity, and degradation of the hydrological cycles, are globally affecting the provision of ecosystem services (MEA 2005).

Ecosystem G&S, or ecosystem functions in general, were also claimed as a candidate for a unifying concept for ecology and economics (de Groot 1987). Ecosystem functions are defined as the capacity of natural processes to provide goods and services that satisfy human needs, directly or indirectly: so, they are "reconceptualized" as ecosystem goods or services when human values are implied (de Groot et al. 2002). The Millennium Ecosystem Assessment (MEA 2005) adopted the following G&S categories:

- supporting services (they support the provision of the other services): nutrient cycling, soil formation, primary production, etc.;
- provisioning services: food, fresh water, wood and fibre, fuel, etc.;
- regulating services: climate, flood, and disease regulation, water purification, etc.;
- cultural services: aesthetic, spiritual, educational, recreational, etc.

These G&S are linked to human well-being and security: the basic material for life, health, social relations, etc. (MEA 2005).

The analysis of ecosystem G&S is increasingly seen as an effective way of integrating relevant indicators to map the state and extent of LDD (Cherlet and Sommer 2009; Sommer et al. 2011). Specified key indicators can ideally reflect human–environment interactions and the associated ecosystem exploitation; hence their variation can be directly related to core ecosystem services, and, in most cases, the assessment of a specific ecosystem service will be a function of the combination and/or integration of several key indicators (Sommer et al. 2011; Zucca et al. 2012).

This approach emphasizes the role of soil information in LDD mapping. Soil is a very dynamic system that performs many functions and delivers services vital to human activities and to the survival of ecosystems.

The proposed EU Soil Framework Directive (European Commission 2006) establishes a framework for the protection of soil and the preservation of the capacity of soil to perform any of the following environmental, economic, social, and cultural G&S:

(a) biomass production, including in agriculture and forestry;
(b) storing, filtering, and transforming nutrients, substances, and water;
(c) biodiversity pool, such as habitats, species, and genes;
(d) physical and cultural environment for humans and human activities;
(e) source of raw materials;
(f) acting as a carbon pool;
(g) archive of geological and archaeological heritage.

To that end, it lays down measures for the prevention of soil degradation processes, both occurring naturally and caused by a wide range of human activities, which undermine the capacity of a soil to perform those functions. Such

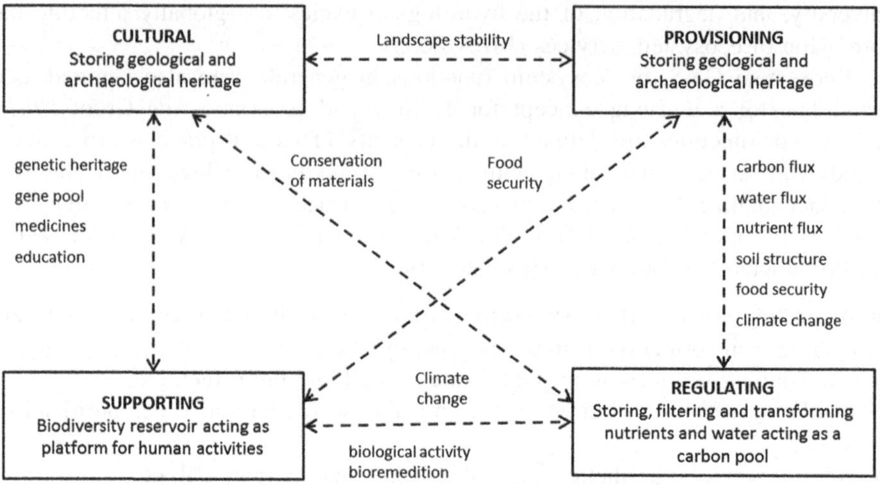

Fig. 3.1 Soil ecosystem services, soil functions, and their interdependencies. *Source* Unpublished material, courtesy of G. Toth

measures include the mitigation of the effects of those processes, and the restoration and remediation of degraded soils to a level of functionality consistent at least with current and approved future use.

The G&S identified in Europe may be as well considered for global assessments; nevertheless, a consistent list is still lacking and should form a core element of the revision of the World Soil Charter (FAO 1981) in the framework of the Global Soil Partnership, as proposed by FAO (2011). A more comprehensive discussion is developed by Toth and Nemeth (2011) and is summarized in Fig. 3.1.

3.2.2 The Anthroscape Concept

The "Anthroscape" concept as defined by Kapur et al. (2004) is a leading candidate for a robust basis for mapping land quality and sustainable land use patterns. This concept, since it embraces the components of the integrated environment, bears significance in assessing human-induced land degradation: understanding soil–landscape relationships in Anthroscapes helps to address LDD, especially when marked differences or deviations from the normal, natural landscapes are observed (Eswaran et al. 2011). Anthroscapes are the result of human impact on natural land- and soilscapes and are therefore typical of the Anthropocene (Crutzen 2002).

The approach is intended to help develop sustainable land management programmes and the introduction of the '*Anthroscape Land Quality Classes*–ALQC' to substitute the classical '*Land Capability Classes*–LCC' of land use of the US Department of Agriculture (Helms 1992).

The current LCC includes eight classes of land designated by the Roman numerals I to VIII. The first four classes are arable land (suitable for cropland) in which the limitations on their use and necessity of conservation measures increase from I to IV, depending on landscape location, slope of the field, depth, texture, and reaction of the soil. The remaining four classes may have uses for pasture, range, woodland, grazing, wildlife, and recreation. Subclasses signify limitations such as (e) erosion, (w) excess wetness, (s) problems in the rooting zone, and (c) climatic limitations. To designate classes not suited to continuous cultivation, the planners typically seize on classes from VI to VIII and subclasses IIIe and IVe. The question is whether the land capability classes, especially IIIe and IVe, are accurate and are a reliable method of identifying erodible land.

The "Anthroscape" context is based on a broader understanding than the LCC, and concentrates on the major issues related to soil loss at basin-wide scale and, thus, requires the integration of the baseline information concerning the topographic, vegetative, land use, demographic and socio-economic attributes with the information on traditional technologies and past land use, with a view to a holistic *sustainable land and water management* (SLWM) programme.

Understanding the soil–landscape relationships in Anthroscapes is a prerequisite for addressing land degradation and desertification. Conventional descriptions and analyses of soils may not suffice to address the subtle changes in soil attributes and functions. The ultimate outcome of the Anthroscape approach is the development of an 'Anthroscape Land Quality Class' map and the relevant 'Ideal Land Use Patterns'. These products are sought as a means of revealing the magnitude and the distribution of the degradation of the selected area, as well as allocating the ideal land use types. In a basin-wide scale, the downstream part of the map would show, so to say, the degradation arising via the intensive cultivation practices, where the class stated in the map would reveal the need of an integrated SLWM programme to revert the lower ALQCs to higher and more sustainable ALQCs. The *net primary productivity* (NPP) can be used via land cover and management as a supplementary indicator of the Anthroscape Land Quality.

3.3 Some Complex Aspects in LDD Assessment and Mapping

LDD takes on a multitude of complex forms and processes in each of the affected regions (Geist 2005) and is highly context-specific (Warren 2002). It can be difficult to take general decisions on which features are to be considered and mapped as LDD indicators, especially if the selected indicators must conserve their validity (and univocal interpretation) over wide areas (e.g. regional to global). This is a major factor explaining why the methodological debate on indicator selection has still not been resolved by the UNCCD (DSD 2009; Orr 2011).

Ecosystem *goods and services* (G&S) provisions are not necessarily evolving in the same direction and some may well improve while others decline. A simple example to illustrate this is the cutting down of a forest to make way for a road because of the needs of a developing economy. Such an intervention triggers a near-immediate decline of the ecosystem G&S. The biomass and a carbon sink disappear overnight, the soil is completely sealed or heavily disturbed, and ecosystem connectivity is affected. In the long run, the presence of this road may become an even more serious ecological threat, favouring penetration of human activities into the forest area, which can become detrimental to the integrity of the ecosystem. Economically, however, the land value is increased through an increased accessibility to markets and possible opportunities for the economic exploitation of the forest areas. The recreational use (and functions) of the forests may also be increased.

The above example also raises the question of the diversity of the G&S and for whom their provision is important. This means that it is necessary to take into account the interests of particular parties that may influence the definition of the concept of land degradation at any one particular time. When LLD occurs, it is often a matter of more concern to some groups of stakeholders than to others, including future generations whose opinion cannot yet be heard.

G&S provided by the land ecosystems also need to be operationally defined, in a way that allows quantification and integration/trade-off between very heterogeneous assets, from economic goods to inspirational services.

LDD is a dynamic process; it involves a change over time in the functioning of the land ecosystem due to human or natural pressures. In this process the land changes from a given state to another, where a decline in functioning is considered as degradation. This implies that a timescale must be set to evaluate LDD, as well as a benchmark baseline against which the changes can be compared.

There is no agreement on the timescale or the baseline year. A pragmatic approach could be to refer to the time span scientifically documented in terms remote sensing and climatic data, about 40 to 50 years maximum. The GLASOD (*Global Assessment of Soil Degradation*) project Oldeman et al. (1990) made an assessment covering the 25 years up to 1990, presumably based on human memory.

Another question is related to the sometimes difficult distinction between ongoing land degradation and degradation inherited from the past. Land degraded in the past may be stable now if pressure factors are no longer active. Furthermore, the productivity of some soils is naturally constrained, as for the natural saline soils. Confusing these stable "bad lands/problem soils" with land presently undergoing bad management and affected by degradation processes would be misleading.

It is often difficult to obtain evidence that past conditions were better than present and that presumed degradation was due to recent or present land use.

Finally, concerning the new definitions proposed for desertification, conceptual and practical problems arise from expressions such as "end state of the process of degradation" and "persistent reduction" because it is problematic to assign them an objective meaning. The same can be said for the distinction between reversible and irreversible conditions, although some authors still claim that what distinguishes desertification is its irreversibility (Santini et al. 2010).

3.4 The Role of Soil Information in Past LDD Mapping

3.4.1 Global Mapping

Mapping land degradation and desertification has generally been considered as a complex task. Past efforts have produced contrasting results, due also to evolving assumptions and methods. The never-ending debate on definitions, along with the scarcity of suitable global datasets, make LDD mapping a real conceptual and operational challenge (Sommer et al. 2011; Zucca et al. 2011).

Several global maps have been drawn since the 1970s, as reviewed by Grainger (2009). Most of the LDD maps produced so far have also been based on expert knowledge (especially mapping of LDD status), or on simple empirical models (mostly adopted to map LDD sensitivity). Among these we can mention:

- the Global Map of Desertification Status (Dregne 1977, 1983);
- the FAO/UNEP Provisional Methodology for Assessment and Mapping of Desertification (FAO/UNEP 1984);
- the Global Assessment of Soil Degradation (GLASOD; Oldeman et al. 1990);
- the UNEP World Atlas of Desertification (UNEP 1992, 1997);
- the Global Desertification Vulnerability Map (USDA 2003).

Apart from the FAO/UNEP (1984) Methodology that mixes status and risk factors, and the US Department of Agriculture (2003) Vulnerability Map, the others are mostly based on expert evaluations of current LDD status. These expert judgements are then usually "attached" to different kinds of predefined map units (either pre-existing or purposely created by overlaying sets of maps). The approaches adopted are most often oriented to reporting on the occurrence of land degradation processes, with a special emphasis on soil degradation. The degradation processes and the related indicators considered by some of these studies are summarized in Table 3.2.

Until recently, GLASOD (Oldeman et al. 1990) was the only global study to estimate the extent of human-induced soil degradation (Fig. 3.2). It discovered that out of the 11.5 billion hectares of vegetated land on Earth, 17 % was degraded largely by erosion, and on this land one in six hectares could no longer be cultivated. The main causes of this environmental disaster were deforestation and adverse farming practices such as overgrazing and nutrient mining. GLASOD estimates are mostly based on expert assessment, but it deserves credit because it contributed greatly to bringing the issue of soil degradation to higher political and decision-making levels.

Other studies that could be mentioned for the inputs and the results they produced at regional scale are the *Assessment of human-induced soil degradation in South and South-East Asia* (ASSOD) (van Lynden and Oldeman 1997), *Digital Geo-referenced database of soil degradation in Russia* (Stolbovoy and Fischer 1997), and *Soil Degradation in Central and Eastern Europe*—SOVEUR (van Lynden 2000).

Table 3.2 Indicators taken into account by some global LDD mapping studies. *Source* The authors

	Processes and factors assessed	Dregne (1977)[1]	FAO/UNEP (1984)[2]	Oldeman et al. (1990)[3]	UNEP (1992)[4]
Soil	Water erosion			X	
	Erosion occurrence				X
	Gullies (number and shape)	X			
	Gully area		X		
	Surface status (stones, rocks…)		X		
	Soil thickness		X		
	Loss of soil depth over root-inhibiting layer		X		
	Eroded area/subsoil exposed		X		
	Wind erosion		X	X	
	Dunes (size, vegetation cover, mobility	X			
	Chemical degradation			X	
	Organic matter reduction		X		
	Salinity	X	X		
	Physical degradation			X	
	Soil crusting and compaction	X	X		
Vegetation	Vegetation degradation				
	Undesired shrubs	X			
	Fall in cover and composition		X		
Yield/ productivity	Farm yield decrease	X			X
	Terrain suitability to local farming				X
	Ease of restoring yields				X
	Ease of restoring terrain				X
Ecosystem	Biotic functions				X

[1] Simple indicators related to four processes (e.g.: "many large, deep gullies"; "undesired shrubs") are assessed in three broad land use classes (rangelands, rain-fed croplands, and irrigated croplands) to define four degrees of desertification severity

[2] Four degrees of desertification severity are assessed based on the evaluation of status, rate, and inherent risk related to six processes. Due to the number of indicators considered, the table only reports the status factors related to water erosion. For this process, the rate is related to the decrease in annual biomass production, the increase in eroded area and soil loss, the sediment deposition in reservoirs, and the related loss of storage, while risk is linked to the climatic aggressivity, the pedo-topographical conditions, and the potential soil erosion

[3] Degree and extent of degradation are assessed for each soil degradation process (four categories, each including several specific indicators) in each of a predefined set of mapping units, to estimate four degrees of desertification severity

[4] Five degrees of desertification severity are considered. Land degradation, and in particular soil degradation, is assessed in two land use systems (rangelands and rain-fed croplands) based on five indicators. These are mostly related to the LDD impact on land functions and health, and on the recovery potential

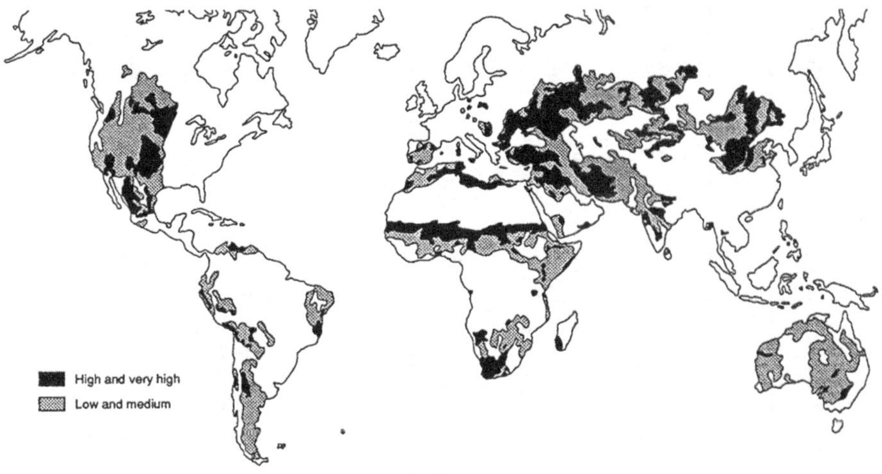

Fig. 3.2 GLASOD, global extent of human-induced soil degradation. *Source* GLASOD

Based on GLASOD, Lal (2003) presented soil degradation data on continental bases and by type of degradation processes (Table 3.3).

3.4.2 Regional to Local Scale Mapping Through Empirical Models

Empirical models have been widely used to combine indicators to perform LDD mapping at various scales, in particular to assess areas prone to desertification. National risk mapping studies were performed in many countries in the frame of the UNCCD implementation, as reviewed for the Northern Mediterranean region (Table 3.4) by various authors in Enne et al (2004), and worldwide by Enne and

Table 3.3 Estimates (millions of ha) of the global extent of soil degradation by different processes. *Source* GLASOD

Region	Total land area	Total degraded area	Total degraded area (%)	Water erosion	Wind erosion	Physical degradation	Chemical degradation
Africa	2,964	494	17	227	186	19	62
Asia	3,085	749	24	441	222	12	74
S.America	1,753	243	14	123	42	8	70
C. America	108	63	58	46	5	5	7
N. America	2,029	96	5	60	35	1	–
Europe	2,260	218	10	114	42	36	26
Oceania	849	102	12	83	16	2	1
World	13,048	1,965	15	1,094	548	83	240

Table 3.4 The indicators considered by the national risk mapping studies performed in the frame of the UNCCD implementation by Northern Mediterranean countries between the late 1990s and the early 2000s. *Source* The authors

Scale	Portugal 1:1,000,000	Spain 1:1,000,000	Italy 1:1,250,000	Greece 1:1,000,000	Turkey 1:1,000,000
Climate	Aridity index P/PET	Aridity index P/PET	Aridity index P/PET	Bioclimatic zones	Annual rainfall; Aridity index P/PET
Soil	Potential soil loss based on four indexes (rainfall erosivity, slope, land use and soil type)	Potential soil loss due to erosion based on *Universal Soil Loss Equation* (USLE) approach	Soil moisture regime	Soil depth and erosion risk based on soil map units and slope; soil salinity based on soil map units and bioclimatic zones	Soil moisture deficiency (Thornthwaite water balance); soil loss (% of topsoil lost)
Vegetation			Land cover classes	Vegetation resilience in bioclimatic zones	NDVI-derived vegetation cover
Drought	Frequency and intensity of drought			Edaphic drought risk related to soil map units and bioclimatic zones	Annual rainfall variability
Demography			Decadal demographic change 1981/1991		
Acquifers over-exploitation		Acquifer pumping and recharge ratio		Irrigation intensity and salt seawater intrusion	
Forest fires		Wildfires occurrence 1986–1995			Fire risk estimated based on fire occurrence per land unit

Yeroyanni (2006) and Begni et al (2007). An attempt was made to integrate the indicators listed in Table 3.4 (with the exception of Turkey) into an overall desertification risk index for all the countries, based on various empirical approaches. In Table 3.4, qualitative, expert-based indicators still coexist with quantitative (although simplified) indicators, but the latter dominate at this spatial scale.

In 2008, the European Environment Agency also produced a map at a scale of 1:1,000,000 showing the sensitivity to desertification for the Northern Mediterranean using a composite evaluation of soil, climate, and relief characteristics, but no socio-economic analyses were considered (EEA 2008).

More sophisticated empirical models were developed to map desertification risk at the local scale. The ESA-Medalus model (Kosmas et al. 1999) is the most widely applied and proved to be flexible enough to be adapted to many different situations (Sepehr et al. 2007; Santini et al. 2010. The soil indicators considered by that model are partly quantitative and partly qualitative: internal soil drainage; texture class; soil depth; rock fragments at the soil surface. These are combined with slope gradient and bedrock type to produce a "soil quality index" by means of map algebra procedures implemented in a *geographical information system* (GIS) environment.

3.4.3 Approaches Based on Vegetation Cover Status as Driving Variable

Some recent efforts to define land status at a small geographical scale have been based on a single index or only a few, such as *Normalized Difference Vegetation Index* (NDVI) (Helldén and Tottrup 2008), NPP (Boer and Puigdefabregas 2005), *rain-use efficiency* (RUE) (Bai et al. 2008), and other statistically more complex methods (Wessel et al. 2008; Del Barrio 2010). These approaches have demonstrated that the interpretation of the results obtained may depend on assumptions that do not conserve their validity over large geographical areas.

A feature common to these methods is that they primarily link LDD trends to vegetation status trend, and do not explicitly consider soil conditions.

3.5 The LADA Project

Since 2006, the *Land Degradation Assessment in Drylands* (LADA[2]) project has been developing and testing a new method for mapping land degradation and land improvement on the basis of a mix of factual data and expert knowledge.

[2] See at: <http://www.fao.org/nr/lada/index.php>.

The method has been conceived and tested at national level in six countries (Argentina, China, Cuba, Senegal, South Africa, Tunisia), covering an area of about seven million square kilometres. It is based on the principle that land use is the main driver of land degradation or land improvement. The method consists of two main steps: (1) the creation of a land use systems map, and (2) the collection of information on the status and trends of land degradation within each of these units.

3.5.1 The Land Use Systems Map

Land use system (LUS) units, according to the LADA methodology (Nachtergaele and Petri 2008), are based on a land cover base with the addition of information on irrigation, livestock, and protected areas. Additional information can be utilized, if available, for further refining the map. This extra information can be attached as an attribute to the unit, in order to have a set of data that allows for a better identification and enhanced explanation of the spatial distribution and the reasons for the land degradation and land improvement. The LUS units are then overlaid with a map of administrative units of the country. The final result consists of LUS units within administrative boundaries. These are the cartographic units for the mapping of land degradation and land improvement at subnational level.

The LADA project has created an LUS map at global level (Fig. 3.3). Each of the participating countries also produced their own national LUS map.

3.5.2 The Collection of Information on Land Degradation and Land Improvement

On the basis of the LUS map, a panel of experts in the country collects and summarizes the available information and knowledge on land degradation and land improvement in each of the LUS units. The method encourages the national panel to collect as much existing hard information as possible in terms of maps, datasets, and statistics. This land degradation information is then attached to the LUS map. This is done through the application of a questionnaire for each cartographic unit (Liniger et al. 2013). The questionnaire includes more than seventy indicators or descriptors of status, causes, and trends of land degradation and improvement status and processes. It is divided into four parts: (1) analysis of the land use system itself, (2) the types, causes, and impacts of land degradation, (3) the measures for combating degradation that are in place, and (4) the recommendations that the evaluation panel proposes for further improvement.

The final output is a georeferenced database containing hundreds of pieces of information on the situation of land resources. More than eighty single-factor maps

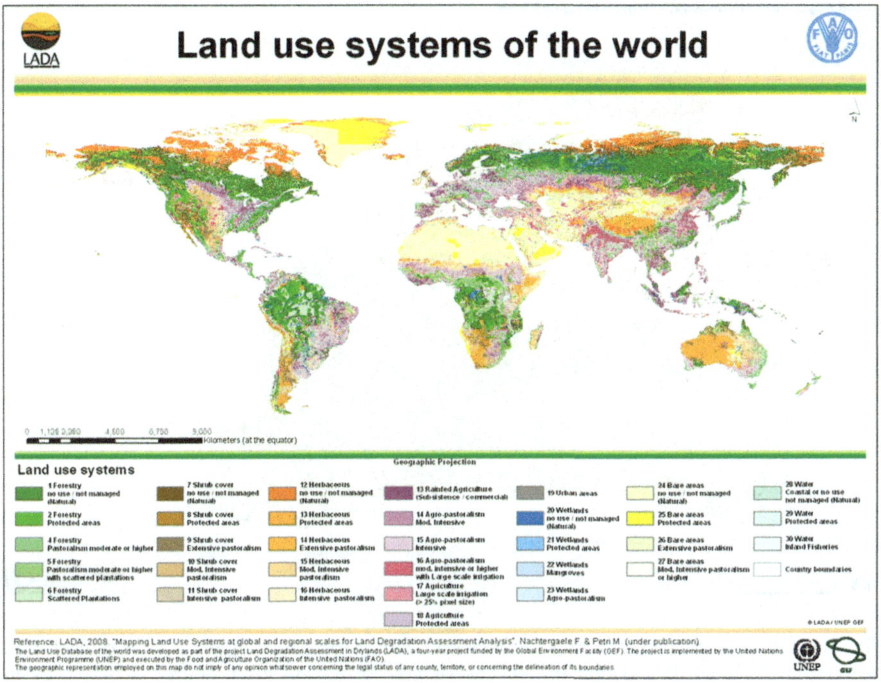

Fig. 3.3 The global Land Use System (LUS) map produced by the LADA project. *Source* Redrawn by the authors, based on unpublished data

can be directly produced from it, giving the possibility of a depiction of the causes and impacts of land degradation at subnational level, according to the *Driver-Pressure-State-Impact and Response* (DPSIR) approach (Gentile 1998). An example of map of land degradation (Land degradation in Senegal—degree of topsoil loss) is given in Fig. 3.4.

3.5.3 Comparison with GLASOD and Role of Soil Information

Although it is recognized that a fully objective assessment of land degradation, based on hard data, is not yet possible (Sonneveld and Dent 2009; Orr 2011), differences do exist between the LADA and GLASOD approaches. Firstly, GLASOD only considers soil degradation, while LADA also takes into account changes in other important ecosystem provisions such as water and vegetation, as well as considering social and economic issues. Secondly, while GLASOD was based on the personal knowledge of a restricted number of experts, LADA relies on panels of specialists from various disciplines, well rooted in their countries,

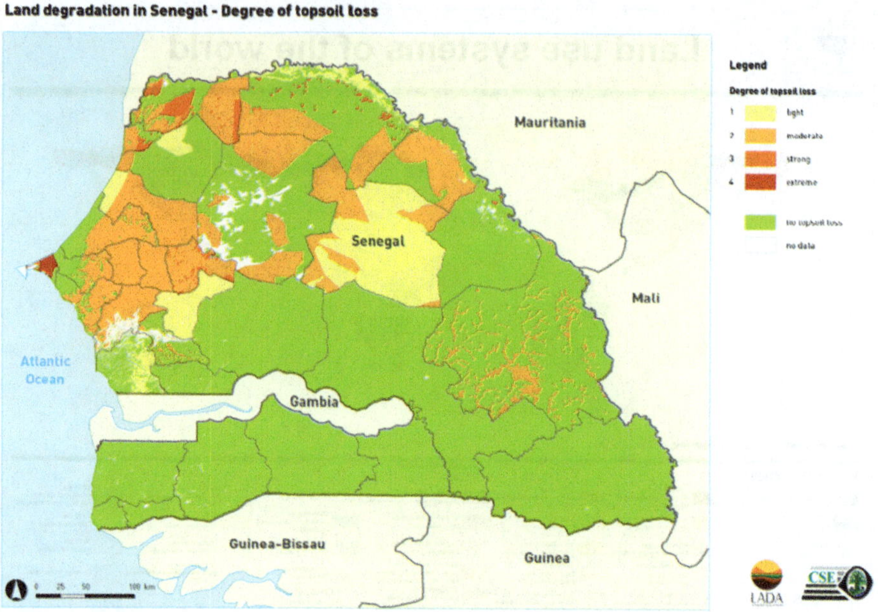

Fig. 3.4 Example of map of a land degradation factor as produced by the LADA project at the national level (Land degradation in Senegal—degree of topsoil loss). *Source* Redrawn by the authors, based on unpublished data

often working at subnational level. The detail and the accuracy of the assessment are therefore greatly increased. Finally, the use of the LUS map and of existing information reinforces the robustness and consistency of the final product.

As stated, LADA considers several aspects of land resources. Nonetheless, soil information still has an important, and sometimes dominant, role in the assessment of the situation and of the possible solution to the problems identified. Soil information is used at two stages of the subnational LADA assessment:

- in the assessment of land degradation, soils are regarded in detail in terms of their physical, chemical, and biological properties and characteristics, which vary from water and wind erosion, salinization, and compaction to alkalinization and many others;
- in terms of pedological information, which is an important attribute that can be attached to the LUS map, allowing a better spatialization of the information on degradation.

3.5.4 Gladis

Parallel to the system for data collection at subnational level described above, the LADA project has developed a system for land degradation assessment at global level, named GLADIS (*Global Land Degradation Information System*; Nachtergaele et al. 2011). GLADIS aims at providing an overview of the status of the land resources and of the processes that act on them, leading either to land degradation or improvement.

Similarly to the subnational system, GLADIS is based on the principle that land use, and in particular land use change, is the main factor leading to land degradation. Hence, a global map of land use systems is also the basis for the interpretation of the outcome in GLADIS. Unlike the subnational system, however, the LUS is not used as the cartographic basis for an expert assessment. On the contrary, in GLADIS the information on natural resources is provided by a series of 32 global datasets, some of them produced by the LADA project, the others collected among several institutions, organizations, and publications. The data have been harmonized into a grid of 5 min of arc.

The information collected in this way is utilized to analyze the status and processes of six main ecosystem "assets", or goods and services: biomass, soil, water, biodiversity, economic, and social. The analysis is done through mostly empirical models, different for the status and for the processes. The models are partially tailored according to the LUS unit, in order to produce results in stronger accordance with the actual reality.

Soil information has a very important role in GLADIS. The status of the soil resources is determined essentially as a soil suitability assessment, on the basis of the actual use of the land. In this way, the risk is avoided of considering certain soils as "good" irrespective of the present cover, so creating the risk of poor decisions based on wrong assumptions. A typical case is the soil under forest, which is usually considered very healthy while the forest remains standing on it, but no assumptions are made where there is a change in land use.

In the GLADIS system, both the chemical (salinization, pollution, and nutrient depletion) and the physical (erosion and compaction) degradation processes are considered. The two groups are then combined to create an index of soil resilience to degradation.

A management index, based on the production performance of the land in agricultural areas, is defined to identify areas potentially undergoing an improvement in the soil conditions.

Two final soil health indexes (status and process) are produced (Fig. 3.5). Together with the other ten indexes (status and process for each of the other five ecosystem assets), they define the overall degradation indexes, again to be analysed on the basis of the land use system units.

3.6 The New World Atlas of Desertification

In response to the interest expressed by the UNCCD *Committee on Science and Technology* (CST) in an updated *World Atlas of Desertification* (WAD), the European Commission's *Joint Research Centre* (JRC), in partnership with UNEP,

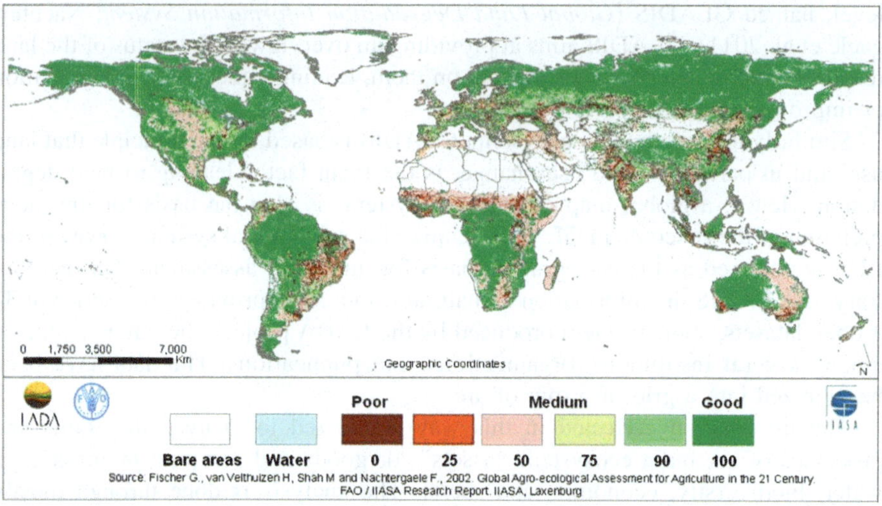

Fig. 3.5 One of the two soil health indexes (status and process) produced by GLADIS: soil health status for present land use. *Source* Nachtergaele et al. (2011). Redrawn by the authors, based on unpublished data

is coordinating the compilation of a new atlas to be ready by the end of 2012 (Cherlet and Sommer 2009; Sommer et al. 2011). The WAD will provide a foundation for addressing the global challenges related to desertification and has the goal of documenting the status of desertification, land degradation, and drought (DLDD) and the factors that influence these processes, and of providing their extent and spatial distribution at various scales: global, continental, regional, and in some cases even at national level. To address the issue of national and especially local scale, the WAD will provide a number of case studies that will be carefully selected to offer worldwide geographical coverage and thematic relevance. The WAD pays particular attention to land use changes and their impacts on ecosystem services.

In the context of WAD preparation, the establishment of a Working Group on "Soil" was proposed at the third Expert Meeting (December 2010). This Working Group will focus on the functioning and impacts of soil and its changing characteristics (constraints and soil degradation), on the human–environment system, and specifically on productivity levels (that is, NPP). Soil constraints and degradation will include a comprehensive assessment and analyses of:

- soil salinization, focusing on impacts related to irrigation;
- water, wind erosion, and sand encroachment, with a major focus on the impacts of unsustainable land use systems reflected in loss of productivity and disruption of ecosystem services;
- organic matter decline and nutrient depletion in arable lands (nutrient mining);
- soil sealing/urbanization as factors in the disruption of water and nutrient cycles affecting both food security and ecosystem stability; and

- soil–water relationships in the drylands, with a major focus on water scarcity as a limiting factor for crop production.

Global maps describing the above transitional processes will be included. The major focus will be on the drylands, so that the intensity of degradation between them and the remaining more humid global areas may be compared.

3.7 The Present Global and Regional Availability of Soil Data

The global situation of soil data availability, be it in the form of soil maps, soil profile information, or chemical, physical, and biological soil characteristics, is unsatisfactory. For too long soils have been seen as a simple substratum in which plants could grow and where water and carbon could be stored. This has resulted in scant investments in soil and soil knowledge over the last forty years. This situation has been further complicated by a number of factors:

- There is no international agreement on how to measure soil properties or on which ones are of prime importance. This has resulted in a multitude of results measured at different times in different ways that cannot easily be compared or correlated.
- Until 1998 there was no international agreement on soil nomenclature. The World Reference Base for Soil Resources established by the IUSS (*International Union of Soil Science*) in 1998 was supposed to end the nomenclature controversy, but it was challenged in several international meetings, making it inapplicable worldwide.
- Soil data are considered valuable and/or private information in many countries and access to national and local soil information can be difficult or expensive to come by.
- There remains a large gap of scientific misunderstanding and economic competition between soil scientists who produce "classical" soil maps using polygons to represent soil associations and those who use "modern" approaches that focus only on point information about certain soil properties.

3.7.1 Availability of Global Soil Maps and Databases

At the global level, the 1:5,000,000 scale Soil Map of the World (FAO-UNESCO 1971–1981) was, until recently, nearly thirty years after its finalization, the only worldwide, consistent, harmonized soil inventory that was readily available in digital format and came with a set of estimated soil properties for each mapping unit (FAO 1995). The digital raster version of this map had a resolution with a $5' \times 5'$ cell size (9 km $\times$ 9 km at the equator), and contained a full database corresponding to the information in the paper map in terms of composition of the soil units, topsoil texture, slope class, and soil phase in each of the more than 5,000 mapping units. In addition pedo-transfer functions allowed further characterization of each unit in terms of chemical and physical properties. This product has been updated over the years, mainly by regional and national efforts under the SOTER (*SOil and TERrain Database*—Van Engelen and Wen 1995) programme run by *International Soil Reference and Information Centre* (ISRIC) and FAO, the production of large regional databases for Europe and the northern circumpolar areas driven by the EU, Russia, and the USA, and efforts by large countries to produce national soil maps, as carried out by China. These updates were brought together in 2008 in the *Harmonized World Soil Database* (HWSD) that combines the recently collected regional and national updates of soil information with the information already contained within the 1:5,000,000 scale FAO-UNESCO Digital Soil Map of the World (FAO/IIASA/ISRIC/ISSCAS/JRC 2008). In order to estimate soil properties in a harmonized way, in this product the use of actual soil profile data and the development of pedo-transfer rules was undertaken in cooperation with ISRIC and the *European Soil Bureau Network* (ESBN), drawing on the ISRIC-WISE soil profile database and earlier work of Batjes et al. (1997, 2007) and Van Ranst et al. (1995). A resolution of about 1 km (30 s of arc by 30 s of arc) was selected. Over 15,000 different soil mapping units are recognized in the HWSD.

The resulting raster database consists of 21,600 rows and 43,200 columns, which are linked to harmonized attribute data. The use of a standardized structure allows linkage of the attribute data with GIS to display or query the map unit composition in terms of soil units and the characterization of selected soil parameters (organic Carbon, pH, soil moisture storage capacity, soil depth, cation exchange capacity of the soil and the clay fraction, total exchangeable nutrients, lime and gypsum contents, sodium exchange percentage, salinity, textural class, and granulometry). The HWSD map, database, and a viewer are available on CD-ROM and freely downloadable from <http://www.fao.org/nr/land/soils/harmonized-world-soil-database/en/>.

The main advantage of these two products is that they are readily and freely available and provide sufficient soil information for global purposes that require information on soil fertility levels, water-holding capacity status (for instance required for Global Circulation Models), or Carbon stocks. Disadvantages are that:

- the reliability of the information presented on the HWSD is variable and depends on the scale/resolution of the source data;
- the approach both in the *Soil Map of the World* (SMW) and the HWSD remains polygon-based and is therefore difficult to translate into point information within the polygon;
- data cannot be used for monitoring purposes and should not be used as a basis for local development plans.

3.7.2 Harmonized Global Soil Profile Databases

In the early 1990s, ISRIC developed a uniform methodology for a global soil database in the framework of a project entitled *World Inventory of Soil Emission Potentials* (ISRIC-WISE). WISE was especially conceived for a geographical quantification of main soil factors that control processes of global change at a broad scale (Batjes and Bridges 1994; Batjes et al. 1995).

In the process of collating materials for compilation of the ISRIC-WISE profile database the quality and validity of the original data had to be evaluated carefully, while at the same time recognizing that these are the only materials available. The description status of the various profiles has been documented in WISE to provide a coarse indicator for the inferred reliability of the source data (Table 3.5).

The latest public domain release of WISE, version 3.1 (WISE3), comprises data for some 10,250 profiles with some 47,800 horizons, from 149 countries (Batjes 2008, 2009), as opposed to around 4350 profiles in an earlier version (Batjes 1999). Most profiles are from Africa (41 %), followed by Asia (18 %), South America (18 %), and Europe (13 %). Their approximate location is shown in Fig. 3.6.

Overall, chemical and physical analyses have taken place in at least 150 laboratories worldwide, using a range of methods; these are described in broad terms in WISE3. Analytical methods used in the *ISRIC Soil Information System* (ISIS) and the *National Soil Survey Center* (NSSC) collection may be considered to be similar (van Reeuwijk 1983). Conversely, for the other sources methods typically vary from one laboratory to the next, even within one country, and may change over time within a single laboratory. These methodological differences complicate the worldwide comparison of soil analytical data, and no single solution for addressing this issue has been found as yet (Batjes et al. 1997; van Reeuwijk 1983; Van Ranst et al. 1995). As a result, the amount of measured data available for modelling is much less than expected.

Table 3.5 Number of profiles in WISE3 by continent and their description status. *Source* Batjes (2009). Reproduced by permission of John Wiley and Sons

Continent	Profile description status[1]				Total
	1	2	3	4	
Africa	421	1,337	2,392	23	4,173
Asia	441	970	426	10	1,847
Antarctica	4	6	0	0	10
Europe	225	712	359	20	1,316
North America	495	222	127	11	855
Oceania	50	49	106	4	209
South America	149	1380	313	1	1,843
Total	1,785	4,676	3,723	69	10,253

[1] The number code under profile description status refers to the completeness and apparent reliability of the soil profile descriptions and accompanying analytical data for the specified profile in the original source; the status is highest for 1 and lowest for 4 (see FAO 2006). Continents are defined according to Times Atlas (2003)

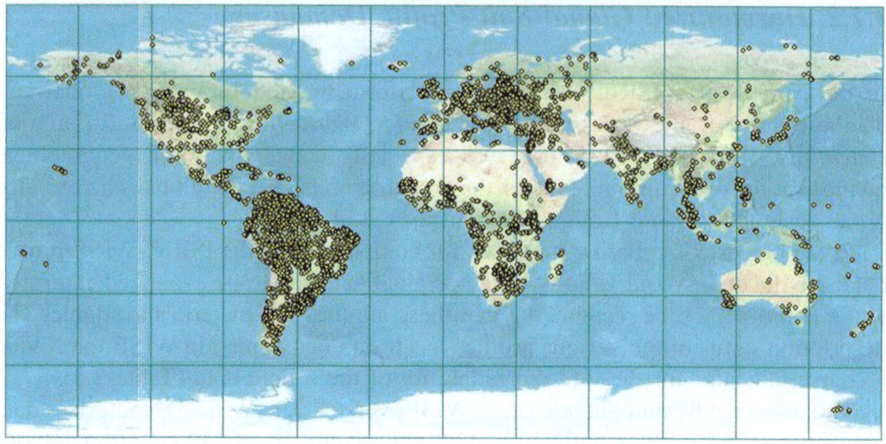

Fig. 3.6 Global distribution of georeferenced soil profiles in WISE3. *Source* Batjes (2008). Reproduced by permission of ISRIC—World Soil Information

The WISE3 dataset is readily and freely available and serves as the main source data both for polygon maps/databases as discussed in Sect. 3.7.1 and for the digital soil mapping approaches discussed in Sect. 3.7.3. Given the uneven distribution in space and time of the profiles and the diversity of analytical methods used, these data cannot be considered independently as a harmonized global product of soil information.

3.7.3 Digital Soil Mapping

In 2006 a consortium of scientific institutes and universities launched an appeal to use the latest satellite technology and new information layers, such as the recently released topographical information at 90-metre resolution, to achieve a spatial database of soil properties based on a statistical sample of landscapes (GlobalSoilMap.net 2009). Within the sample of satellite sites, field sampling is used to determine the spatial distribution of soil properties in order to develop reflectance spectral libraries for the characterization of soil properties (Shepherd and Walsh 2002). These are then used to predict soil properties in areas not sampled (see McBratney et al. 2003; Lagacherie et al. 2006; Hartemink et al. 2008). The resulting digital soil maps describe the uncertainties associated with such predictions and, when based on time series data, can also provide information on dynamic soil properties. Maps derived from digital soil mapping differ from conventional polygon-based maps in that they are pixel-based and thus can be more easily displayed at a higher resolution than those currently used by other earth and social sciences (Sanchez et al. 2009).

The GlobalSoilMap.net (2009) project, funded by the Bill & Melinda Gates Foundation in 2008, started with a pilot covering sub-Saharan Africa. Regional nodes in other continents were also established. As summarized by Sanchez et al. (2009), GlobalSoilMap.net proceeds in steps with the ultimate aim of developing evidence-based soil management recommendations at very high resolution. During the initial stages of the project, only six soil properties are considered: clay content, organic carbon content, pH, estimated cation exchange capacity, electrical conductivity, and bulk density. These may be used to generate a range of maps, as discussed by Sanchez et al. (2009). Finally, spatially inferred soil properties are used to predict more difficult-to-measure soil functions such as available soil water storage, carbon density, and phosphorus fixation. This is achieved by using pedotransfer functions.

An innovative element of the approach is that the overall uncertainty of the prediction is determined by combining uncertainties of the input data, a spatial inference model, and the soil functions used.

The approach relies heavily on statistics and modern approaches to soil mapping (spectrophotometry, remote sensing), limiting the use of expensive systematic ground observations. It moves away from understanding the soil distribution in a landscape to achieving a statistical representation of soil properties for soil management purposes. Several pilot projects are under way to support the theories that underlie GlobalSoilMap. Conversely, as with any new approach, a number of scientific and operational challenges still need to be resolved; these have been discussed in detail by various authors (Lagacherie et al. 2006; Hartemink et al. 2008).

3.8 Data Gap and Needs and Recommendations for Global Soil Data Gathering, Survey, Processing, and Use

As illustrated in the previous section, there are significant gaps in global and regional soil information in terms of uniformity and harmonization. Furthermore, the coverage is uneven, with drylands, deserts, mountains, and Polar Regions having very few measured data, while regularly monitored soil data are lacking nearly everywhere. Resources for making new soil inventories are scarce and many national soil survey agencies have closed down in recent years. There are additional problems with making the soil information freely available in an easily accessible format. Problems are as much economic and political as they are scientific.

Overall there appears to be an urgent need to come to a binding international agreement on soil nomenclature and soil laboratory methods. This would need to be supplemented by accepting the complementarities of polygon-based and point-based approaches to classical or digital soil mapping as presently undertaken in the e-SOTER, DIGISOIL and iSOIL projects of the European Union. Due strengthening of national soil agencies and their full involvement in international initiatives is also a must.

This can only be achieved by a new agreement between all parties, as recently proposed under the *Global Soil Partnership* (GSP), that would bring into a coherent framework all current data collection and soil mapping efforts at the global scale. The main elements of the GSP, as proposed by FAO, will be a vigorous effort towards standardization and harmonization of data collection and soil mapping methodologies that combines the existing raster-based and polygon-based digital soil mapping projects into a single coherent framework. An essential part of this renewed effort will be the adoption of a new Universal Soil Classification system that will combine existing national and international systems into a single unified global standard.

3.9 Conclusions

Most of the global LDD status maps produced since the 1970s were aimed at reporting on the occurrence of degradation processes within predefined map units. Soil degradation was generally given a special emphasis, but the maps were mainly qualitative and based on expert knowledge. Some mapping exercises at greater scales, such as the studies performed by countries in the framework of the UNCCD, were mainly based on empirical models (especially the mapping of LDD risk) where qualitative and quantitative soil data coexisted.

Since most of these methods were oriented towards tracking processes, little attention was devoted to causes and to impacts on ecosystem G&S.

The most recent approaches (especially LADA and WAD) try to overcome these limitations. Both initiatives integrate innovative approaches to interlink causal factors and to represent the impacts on ecosystem G&S. LADA designed a sound conceptual framework for integrating qualitative and quantitative information. WAD will be based on hard data and will make full use of the new global soil datasets.

The degree of harmonization and availability of global soil data has considerably improved during the last few years. The development of pedo-transfer functions is helping to fill gaps in information, but there is an urgent need for updated and detailed data and information on policy-relevant soil parameters, particularly soil organic carbon content, soil erosion, salinization, contamination, compaction, soil biodiversity levels, and others. Unfortunately ongoing efforts at data collection are very limited and lack the necessary multi disciplinarity for addressing the policy-relevant issues at stake. In addition, competing initiatives and lack of standardization are generating unnecessary duplication of efforts and a waste of available resources. The new Global Soil Partnership started in 2012, with Secretariat at FAO could be the way forward towards the next generation of policy-relevant soil data and information at the global scale.

References

Bai, Z.G., Dent, D.L., Olsson, L., Schaepman, M.E., 2008: "Proxy global assessment of land degradation", in: *Soil Use Management,* 24: 223–234.

Batjes, N.H., 1999: "Development of a 0.50 by 0.50 resolution global soil database". In: Sumner, M. (Ed.): *Handbook of Soil Science* (Boca Ratton: CRC Press): 29–40 (Chap. H).

Batjes, N.H., Al-Adamat R., Bhattacharyya T., Bernoux M., Cerri C.E.P., Gicheru P., Kamoni P., Milne E., Pal D.K., Rawajfih Z., 2007: "Preparation of Consistent Soil Data Sets for SOC Modelling Purposes: Secondary SOTER Data Sets for Four Case Study Areas", in: *Agriculture, Ecosystems and Environment*, 122: 26–34.

Batjes, N.H., 2008: "ISRIC-WISE Harmonized Global Soil Profile Dataset (Version 3.1)", in: *Report 2008/02 ISRIC—World Soil Information, Wageningen.* <http://www.isric.org/isric/webdocs/Docs/ISRIC_Report_2008_02.pdf>.

Batjes, N.H., 2009: "Harmonized soil profile data for applications at global and continental scales: updates to the WISE database", in: *Soil Use and Management*, 25: 124–127.

Batjes, N.H., Bridges, E.M., 1994: "Potential emissions of radiatively active gases from soil to atmosphere with special reference to methane: development of a global database (WISE)", in: *Journal of Geophysical Research,* 99(D8): 16479–16489.

Batjes, N.H., Bridges, E.M., Nachtergaele, F.O., 1995: "World Inventory of Soil Emission Potentials: Development of a Global Soil Data Base of Process-Controlling Factors", In: Peng S., Ingram K.T., Neue H.U., Ziska L.H., (Eds.): *Climate Change and Rice* (Heidelberg: Springer): 102–115.

Batjes, N.H., Fischer, G., Nachtergaele, F.O., Stolbovoy, V.S., van Velthuizen, H.T., 1997: "Soil Data Derived from WISE for Use in Global and Regional AEZ Studies (ver. 1.0)", in: *Interim Report* IR-97–025, FAO/IIASA/ISRIC, Laxenburg <http://www.iiasa.ac.at/Admin/PUB/Documents/IR-97-025.pdf>.

Begni, G., Enne G., Zanolla C., Zucca C., Siheng S., Klintenberg P. 2007: Echange actif d'experiences sur les indicateurs et le developpement de perspectives dans le domaine de la desertification et le contexte de l'UNCCD: le projet AID-CCD. Revue Française de Photogrammetrie et de Teledetection, n°187-188. Pp 76-83.

Boer, M.M., Puigdefabregas Journal 2005: "Assessment of dryland condition using spatial anomalies of vegetation index value", *International Journal Remote Sensor*, 26: 4045–4065.

Cherlet, M, Sommer, S. 2000: "WAD Implementation Plan. Roadmap Towards a new World Atlas on Desertification, Land Degradation and Drought", in: *Joint Research Centre, Institute for Environment and Sustainability.* Action DESERT (2008), WP 2009, Deliverable 2.2. http://desert.jrc.ec.europa.eu/action/php/index.php?action=view&id=105.

Crutzen, P.J., 2002: "Geology of mankind", in: Nature, Vol. 415:23.

de Groot, R.S., 1987: "Environmental Functions as a Unifying Concept for Ecology and Economics", in: *The environmentalist*, 7: 105–109.

de Groot, R.S., Matthew, Wilson, A., Roelof, M., Boumans, J., 2002: "A typology for the classification, description and valuation of ecosystem functions, goods and services, Ecological Economics", 41: 393–408.

del Barrio, G., Puigdefabregas, J., Sanjuan, M.E, Stellmes, M., Ruiz, A., 2010: "Assessment and monitoring of land condition in the Iberian Peninsula" 1989–2000 In: *Remote Sensor Environment* 114, 1817–1832.

Dregne H.E., 1977: "Map of the Status of Desertification in the Hot Arid Regions. Document" A/CONF.74/31. United Nations Conference on Desertification, Nairobi.

Dregne, H., 1983: *Desertification of arid lands*, (London: Harwood Academic Publisher).

DSD, 2009: "Integrated Methods for Assessment and Monitoring Land Degradation Processes and Drivers (Land Quality)" in: *White paper of the DSD Working Group* 1, *version* 2. http://dsd-consortium.jrc.ec.europa.eu/documents/WG1_White-Paper_Draft-2_20090818.pdf (last accessed Nov 2010).

EEA, 2008: "Mapping Sensitivity to Desertification (DISMED)", in: *Final report, version 2. European Topic Centre on Land Use and Spatial Information and European Environment Agency*. [online] URL: http://www.eea.europa.eu/data-and-maps/data/sensitivity-to-desertification-and-drought-in-europe/mapping-sensitivity-to-desertification-dismed-2008/mapping-sensitivity-to-desertification-dismed-2008 [last accessed 14 April 2011].

Enne, G., Yeroyanni, M., 2006: "AIDCCD—Report on the State of the Art on Existing Indicators and CCD Implementation in the UNCCD Annexes", Sassari, 351 pp.

Enne G., Peter D., Zanolla C., Zucca C., (Eds): 2004: "The MEDRAP Concerted Action to Support the Northern Mediterranean Action Programme to Combat Desertification", *Workshops results and proceedings NRD*. Sassari. 938 pp.

Eswaran, H., 1994: "Soil Resilience and Sustainable Land Management", In: Greenland D.J., Szabolcs I.,(Eds.): *Soil Resilience and Sustainable Land Use, Wallingford,* (UK: CAB International): 21–32.

Eswaran, H., Lal, R., Reich, P.F., 2001: "Land Degradation: an Overview" in: Bridges, E.M., Hannam, I.D., Oldeman, L.R., Pening de Vries, F.W.T., Scherr, S.J., Sompatpanit S., (Eds.): *Response to land degradation*, (Enfield (NH) Science Publishers Inc.,): 20–35 pp.

Eswaran, H., Beinroth, F.H., and Reich, P., 2003: "A global assessment of land quality", In: *Land quality, agricultural productivity, and food security: biophysical processes and economic choices at local, regional, and global levels*, Wiebe K., (Ed.): (Edward Elgar Publishing Ltd) 111–132pp.

Eswaran H., Berberoğlu S., Cangir C., Boyraz D., Zucca C., Özevren E., Yazıcı E., Zdruli P., Dingil M., Dönmez C.,Akça E., Çelik I., Watanabe T., Koca Y.K., Montanarella L., Cherlet M., Kapur S., 2011: "The Anthroscape Approach in Sustainable Land Use". In: Kapur S., Eswaran H., Blum W.E.H., (Eds): *Sustainable Land Management—Learning from the Past for the Future*. (Heidelberg: Springer): ISBN 978-3-642-14781-4. 1–50pp.

European Commission, 2006: "Proposal for a Directive of the European Parliament and of the Council Establishing a Framework for the Protection of Soil and Amending Directive" 2004/35/EC, COM(2006) 232.

FAO, 1995a: "Digital Soil Map of the World and Derived Soil Properties, Land and Water Digital Media Series 1", Food and Agriculture Organization of the United Nations, Rome.

FAO, 2006: "Guidelines for soil description", Fourth Edn: FAO, Rome.

FAO, 2011: "Towards the Establishment of the Global Soil Partnership", in: *Summary report and way forward of the launch event* 7–9 September 2011, FAO Headquarters, Rome, Italy.

FAO/IIASA/ISRIC/ISSCAS/JRC, 2008: "Harmonized World Soil Database (version 1.0), Food and Agriculture Organization of the United Nations", International Institute for Applied Systems Analysis, ISRIC—World Soil Information, Institute of Soil Science—Chinese Academy of Sciences, Joint Research Centre of the European Commission, Laxenburg.

FAO/ISRIC/ISSS, 2006: "World reference base for soil resources 2006", in: *A framework for international classification, correlation and communication*. FAO, World Soil Resources Reports n. 103, Roma.

FAO/UNEP, 1984: "Provisional methodology for assessment and mapping of desertification", FAO, Rome.

FAO-Unesco, 1971–1981: "Soil Map of the World", 1:5,000,000. Vol. 1–10. United Nations Educational, Scientific, and Cultural Organization, Paris.

Geist, H.J., 2005: "The Causes and Progression of Desertification", (Ashgate Publishing Limited, Aldershot, England): ISBN 0-7546-4323-9.

Gentile, A.R., 1998: "From National Monitoring to European Reporting: The EEA Framework for Policy Relevant Environmental Indicators". In Enne G., d'Angelo M., Zanolla C., *Proceedings of the International Seminar on Indicators for Assessing Desertification in the Mediterranean*, Porto Torres (Italy) 18–20 September, 16–26 pp.

GlobalSoilMap.net, 2009: Project overview available at: http://www.globalsoilmap.net/. (last accessed June 2011).

Grainger, A., 2009: "Development of a Baseline Survey for Monitoring Biophysical and Socio-Economic Trends in Desertification, Land Degradation and Drought", in: *Report* UN Convention to Combat Desertification, Bonn.

Hartemink, A.E., Mc Bratney, A., Mendonca-Santos, M., 2008: "Digital Soil Mapping with limted data", (Springer): 600 pp.

Helldén, U., Tottrup, C., 2008: "Regional desertification: A global synthesis" in: *Global Planet Change* 64, 169–176.

Helms, D., 1992: "Readings in the History of the Soil Conservation Service", Washington, DC: Soil Conservation Service, 60–73 pp.

Herrmann, S.M., Hutchinson, C.F., 2005: "The changing contexts of the desertification debate", in: *Journal Arid Environment*, 63, 3. 538–555.

Kaiser, J., 2004: "Wounding Earth's Fragile Skin", in: *Special edition: Soil the Final Frontier.* Science 304:1616–1622.

Kapur, S., Zdruli, P., Akça, E., Arnoldussen, A.H., Kapur, B., 2004: Anthroscapes of Turkey: Sites of Historical Sustainable Land Management (SLM). In: van Asselen S., Boix-Fayos C., Imeson A. (Eds.): *Briefing papers of the second SCAPE Workshop*, Cinque Terre (Italy), 13–15 April.

Kosmas, C., Kirkby, M., Geeson, N., 1999: "The MEDALUS project", in: *Mediterranean Desertification and land use. Manual on key indicators of Desertification and mapping environmentally sensitive areas to desertification*, European Commission, Brussels.

Lagacherie, P., McBratney, A., Voltz, M., (Eds.): 2006: "Digital Soil mapping: An introductory perspective" (Elsevier: Amsterdam): 350 p.

Lal, R., 2003: "Soil Degradation and Global Food Security: a Soil Science Perspective". In: *Land quality, agricultural productivity, and food security: biophysical processes and economic choices at local, regional, and global levels*, Wiebe K., (Ed.): Edward Elgar Publishing Ltd. 16–35pp.

Le Houérou, H.N., 2002: "Man-Made Deserts: Desertization Processes and Threats", in: *Arid Land Resource Management*, 16, 1. 1–36.

Liniger, H., van Lynden, G., Nachtergaele, O.F., Schwilch, G., Biancalani, R., 2013: Questionnaire for Mapping Land Degradation and Sustainable Land Management, FAO, Rome

McBratney, A.B, Mendonça-Santos, M.L., Minasny, B., 2003: "On Digital Soil Mapping", in: *Geoderma*, 117, 3–52.

MEA—Millennium Ecosystem Assessment, 2005: "Ecosystems and human well-being: Desertification Synthesis", (Island Press: Washington DC, USA).

Nachtergaele, O.F, Petri, M., 2008: "Mapping Land Use Systems at Global and Regional Scales for Land Degradation Assessment Analysis", FAO-LADA.

Nachtergaele, O.F., Petri, M., Biancalani, R., van Lynden, G., van Velthuizen, H., 2011: "Global Land Degradation Information System (GLADIS)—Version 1", in: *An information database for Land Degradation Assessment at Global Level.* LADA Technical Report n. 17. http://www.fao.org/nr/lada/index.php?option=com_content&view=article&id=181&Itemid=113&lang=en (last accessed September 2011).

Oldeman, L.R., Hakkeling, R.T.A., Sombroek, W.G., 1990: "World map of the status of human-induced soil degradation", in: *An explanatory note.* ISRIC/UNEP, Wageningen.

Orr, B.J., 2011: "Scientific Review of the UNCCD Provisionally Accepted Set of Impact Indicators to Measure the Implementation of Strategic Objectives 1, 2 and 3", White Paper—Version 1. UNCCD. Bonn. http://www.unccd.int/science/docs/Microsoft%20Word%20-%20White%20paper_Scientific%20review%20set%20of%20indicators_Ver1_31011%E2%80%A6.pdf (accessed February 2011).

Reynolds, J.F., Stafford Smith, D.M., Lambin, E.F., Turner II, B.L., Mortimore, M., Batterbury, S.P.J., Downing, T.E., Dowlatabadi, H., Fernandez, R.J., Herrick, J.E., Huber-Sannwald, E., Jiang, H., Leemans, R., Lynam, T., Maestre, F.T., Ayarza, M., Walker, B., 2007: "Global Desertification: Building a Science for Dryland Development", in: *Science* 316, 847–851.

Safriel, U.N., 2007: "The Assessment of Global Trends in Land Degradation", in: Sivakumar, M.V.N., Ndiang'ui, N., (Eds.): *Climate and land degradation*, (Springer: Berlin): 1–38 pp.
Safriel U.N. 2009: "Status of Desertification in the Mediterranean Region", In: *Water Scarcity, Land Degradation and Desertification in the Mediterranean Region*. Rubio J.L., Safriel U.N., Daussa R., Blum, W.E.H., Pedrazzini F., (Eds): NATO Science for Peace and Security Series C: Environmental Security, Springer Science + Bussines Media B.V. 33–73 pp.
Sanchez, P.A., Ahamed, S., Carré, F., Hartemink, A.E., Hempel, J., Huising, J., Lagacherie, P., McBratney, A.B., McKenzie, N.J., de Lourdes Mendonça-Santos, M., Minasny, B., Montanarella, L., Okoth, P., Palm, C.A., Sachs, J.D., Shepherd, K.D., Vågen, T-G., Vanlauwe, B., Walsh, M.G., Winowiecki, L.A., Zhang, G-L., 2009: in: *Science* 325, 680–681.
Santini, M., Caccamo, G., Laurenti, A., Noce, S., Valentini, R. 2010: "A Multi-Component GIS Framework for Desertification Risk Assessment by an Integrated Index". in: *Applied Geography* 30, 394–415.
Sepehr, A., Hassanli, A.M., Ekhtesasi, M.R., Jamali, J.B., 2007: "Quantitative Assessment of Desertification in South of Iran Using MEDALUS Method", in: *Environment Monit Assess.*
Shepherd, K., Walsh, M., 2002: "Development of Reflectance Spectral Libraries for Characterization of Soil Properties. in: *Soil Science Society of America Journal* 66, 988–998.
Sommer, S., Zucca, C., Grainger, A., Cherlet, M., Zougmore, R., Sokona, Y., Hill, J., Della Peruta, R., Roehrig, J., Wang, G., 2011: "Application of Indicator Systems for Monitoring and Assessment of Desertification from National to Global Scales", in: *Land Degradation and Development*, 22, 184–197.
Sonneveld, B.G.J.S., Dent, D., 2009: "How good is GLASOD?", in: *Journal of Environmental Management* 90, 274–283.
Stolbovoy, V., Fischer, G., 1997: "A New Digital Georeferenced Database of Soil Degradation in Russia", in: *International Institute for Applied Systems Analyses*. Interim Report IR-97-084/November. 14 pages. http://www.iiasa.ac.at/Publications/Documents/IR-97-084.pdf. (last accessed June 2011).
Thomas, D.S.G., Middleton, N.J., 1994: "Desertification; Exploding the Myth", Wiley: Chichester.
Times Atlas, 2003. The Comprehensive Times Atlas of the World. Times Books, Harper Collins Publ., London.
Toth, G., Nemeth, T., 2011: "Land Quality and Land Use Information in the European Union". EUR 24590 EN, ISBN 978-92-79-17601-2, (Publication Office of the European Union: Luxembourg).
UNCCD, 1994: "United Nations Convention to combat Desertification in those countries experiencing serious drought and/or desertification, particularly in Africa", UNEP, Geneve.
UNEP, 1992: "World Atlas of Desertification". Thomas, D.S.G., Middleton, N.J., (Eds.): Arnold, London.
UNEP, 1997: "World Atlas of Desertification", Thomas, D.S.G., Middleton, N.J. (Eds.), 2nd Edn. Arnold, London.
USDA, 2003: "Global Desertification Vulnerability Map", USDA-NRCS, Soil Survey Division, World Soil Resources, Washington D.C. http://soils.usda.gov/use/worldsoils/mapindex/desert.html (last accessed November 2010).
van Engelen, V.W.P., Wen, T.T., 1995: „Global and National Soils and Terrain Digital Databases (SOTER)", Procedures Manual (revised edition), FAO, ISSS, ISRIC, Wageningen.
Van Lynden, G.W.J., 2000: „Soil degradation in Central and Eastern Europe". The assessment of the status of human-induced soil degradation in. FAO and ISRIC, Report 2000/5. 15 p.
Van Lynden, G.W.J., Oldeman, L.R., 1997: "Assessment of the status of human induced soil degradation in South and South East Asia", in: *International Soil Reference and Information Centre*, Wageningen. 35 p. http://www.isric.org/sites/default/files/ASSODEndReport.pdf. (last accessed June 2011).
Van Ranst, E., Vanmechelen, L., Thomasson, A.J., Daroussin, J., Hollis, J.M., Jones, R.J.A., Jamagne, M., King, D., 1995: "Elaboration of an Extended Knowledge Base to Interpret The 1:1 000 000 Eu Soil Map for Environmental Purposes. In: King, D., Jones, R.J.A.,

Thomasson, A.J. (eds). *European Land Information Systems for Agro-environmental Monitoring*. EUR 16232 EN, 71–84 p. (Office for Official Publications of the European Communities: Luxemburg).

van Reeuwijk, L.P., 1983: "On The Way to Improve International Soil Classification And Correlation: The Variability of Soil Analytical Data", in; *Annual Report* 1983, ISRIC, Wageningen. (http://www.isric.org/isric/webdocs/Docs/annual_report_1983.pdf).

Warren, A., 2002: "Desertification is Contextual", in: *Land Degrad.* Dev. 13, 449–459.

Wessels, K.J., Prince, S.D., Reshef, I., 2008: "Mapping Land Degradation By Comparison Of Vegetation Production To Spatially Derived Estimates Of Potential Production", *Journal Arid Environment*, 72, 1940–1949.

Zucca, C., Bautista S., Orr B.J., Previtali F., 2012: "Desertification: Prevention and Restoration". Chapter XX in: Jorgensen S.E., (Ed.): *Encyclopedia of Environmental Management* (EEM). (New York: Taylor & Francis Group, LLC).

Zucca, C., Della Peruta, R., Salvia, R., Cherlet, M., Sommer, S., 2011: "Evaluation and Integration of Baseline Indicators for Assessing and Monitoring Desertification", European Communities, EUR Report in press. ISSN 1018-5593, Luxembourg, Office for Official Publications of the European Communities.

Zucca, C., Della Peruta, R., Salvia, R., Sommer, S., Cherlet, M., 2012: "Towards a World Desertification Atlas", in: *Relating and selecting indicators and datasets to represent complex issues. Ecological Indicators* 15, 157–170.

Chapter 4
Pedoenvironments of the Mediterranean Countries: Resources and Threats

Franco Previtali

Abstract The Mediterranean regions possess, besides a great cultural heritage and an advanced economy and technology, remarkable environmental resources, consisting of a variety of soils, landscapes, waters, fauna, and flora. But high anthropic pressure, global warming, the heterogeneity of state interventions, even urbanization and "littoralization", as well as some rather short-sighted policies, are jeopardizing the quality and conservation of many ecosystems and the productive capacities of soils. In this paper the peculiar properties of the Mediterranean soils and the degradation and consumption processes closely related to anthropic activities and the cycles of erosion are briefly examined.

Keywords Soils · Mediterranean countries · Environment · Resources · Threats

4.1 Introduction

The countries bordering the Mediterranean Sea possess—besides cultures and traditions that have influenced the history, thought, and value systems of many peoples on many continents—a vast heritage of natural resources. A large number of these resources are directly or indirectly dependent on some of the basic features of these regions: *climate*, *soils*, and *population pressure*.

It is common knowledge that climatic conditions are seriously threatened by global change, and this seems to trigger extreme meteorological events, and in particular alternating floods and drought that occur with alarming frequency in regions not usually involved with such phenomena. It appears that the development model followed in the industrialized countries—and unfortunately imitated

F. Previtali (✉)
Department of Earth and Environmental Sciences, University of Milano-Bicocca, Piazza della Scienza 1 20126 Milano, Italy
e-mail: Franco.previtali@unimib.it

S. Kapur and S. Erşahin (eds.), *Soil Security for Ecosystem Management*, SpringerBriefs in Environment, Security, Development and Peace 8,
DOI: 10.1007/978-3-319-00699-4_4, © The Author(s) 2014

in many developing countries—at least intensifies the effects of global warming and accelerates land degradation (Van Lynden 2000).

As far as soil, an essential natural factor contributing to production, is concerned, the Mediterranean countries (Fig. 4.1) are endangering its quality and productivity through frequently irrational uses and through uncontrolled building, and are failing to consider its intrinsic value and the functions it exerts within ecosystems. It is evident that both rich and poor countries are involved in these negative processes.

This picture is worsened by inadequate knowledge of the quantitative dimension of land degradation, by the diversity of the assessment methods adopted, and by the frequent absence of economic evaluations in surveys and studies of the problem (FAO 1976). Furthermore, the investigations of land degradation are not diachronic, and so lack an efficient multitemporal monitoring. It is obvious that the heterogeneous political framework of the Mediterranean countries makes common management of resources, especially water and soil, particularly random (CIHEAM 1993).

One frequently omitted aspect of environmental analysis carried out at the macrosystemic level is the strong influence exerted on the quality of environmental resources, and consequently on human living conditions, by civil and international wars. Attention is as a rule correctly focused on loss of life and the destruction of dwellings and infrastructure. But the long-lasting contamination of water, soil, vegetation, fauna, and air by depleted uranium, heavy metals, hydrocarbons and so on, and the drastic modification of landforms by bombing, are left out or underestimated.

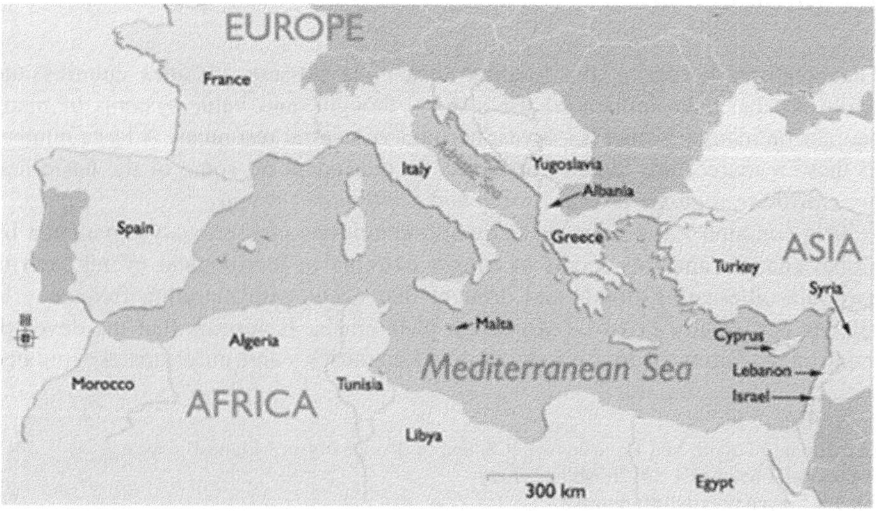

Fig. 4.1 The Mediterranean countries. *Source* Zdruli et al. (2008)

4.2 Geographical Outlines of the Mediterranean Environment

The Mediterranean coastline (Fig. 4.1) is about 46,000 km long, including several islands. In 2000, the coastal zone accommodated 143 million inhabitants in 234 coastal administrative entities (Benoit and Comeau 2005). This means that one-third of the Mediterranean population lives on about one-eighth of the area of the coastal countries. This is the "littoralization" of Zdruli et al. (2008, 2010). Overdevelopment and artificial land cover typify this process, intensified by domestic and foreign tourism. The population in the Mediterranean coastal regions grew from 95 million in 1970 to 143 million in 2000, at an average annual growth rate of 1.4 %. According to Benoit and Comeau (2005), by 2025 this population will reach 174 million inhabitants, at an annual growth rate of 0.8 %.

4.3 Environmental Conditions and Soils

4.3.1 Climate

It is generally known that not only the sea but also the countries that border it and by inference a particular type of climate have been given the name "Mediterranean". This climate is characterized not so much by its total annual rainfall (200–1,000 mm and over) as by its particular seasonal distribution: dry and hot summers and moist and cool winters. The average annual temperatures do not exceed, by definition, 20 °C (Yaalon 1997).

Expressed in terms of soil temperature and soil moisture regimes using the Soil Taxonomy (Soil Survey Staff 1999), the Mediterranean soils have (Verheye and de la Rosa 2005):

- a *xeric* moisture regime, characterized by rainfall occurring immediately after the winter solstice, and followed by a significant dry period after the summer solstice;
- a *thermic* temperature regime (mean annual soil temperature between 15 and 22 °C) or occasionally *mesic* (8–15 °C), that is, intermediate between the temperate regions and the tropics.

4.3.2 Geology

The Alpine orogenesis caused the upthrust of mountain chains around the Mediterranean (Atlas, Betics, Pyrenees, Apennines and Alps, Dinarides, Pindus, Taurus, Lebanon mountains), in places reaching and exceeding an altitude of 4,000 m. Their geologically young age explains their broken and dissected relief, with deep valleys and steep slopes. Slopes exceeding 8 % occupy 45 % of the whole region (de Franchis 2003). Besides this, due to a lower density of vegetation, the south-facing slopes are subject to stronger water erosion (Morgan 2005).

All kind of sedimentary, igneous, and metamorphic rocks are represented in these regions, but the carbonate formations (limestones, dolomites, and marls) seem to be the most common parent materials of the soils in Mediterranean areas, with different silicate content, weatherability, fracturation, and permeability.

4.3.3 Vegetation

In the absence of human intervention, the natural vegetation in the areas bordering the Mediterranean would consist of holm oak (*Quercus ilex*), cork oak (*Q. suber*), wild olive (*Olea europeae*), carob (*Ceratonia siliqua),* lentisk (*Pistacia lentiscus*), stone pine (*Pinus pinea*), and Aleppo pine (*P. halepensis*). Degradation of the original forest often results in the establishment of a thorny xerophytic vegetation, called guarrigue on calcareous soils or maquis on acid soils (Verheye and de la Rosa 2005).

4.3.4 Soil Resources

If we limit the field of observation to the regions bordering the Mediterranean, about one-third of the thirty-two World Reference Base soil groups comprising the soil classification of the IUSS Working Group WRB (2006) can be considered to be absent or at least rarely occurring, being where they occur remnants of pre-Holocene climatic conditions. Groups represented only by extreme and scattered sites are: those from the cold polar regions (restricted to the highest mountains); those typical of highly weathered humid tropics (in this case, palaeosols); and those from the inner continental steppes and grassland.

In more detail, and referring to the World Reference Base soil classification (IUSS Working Group WRB 2006), here is the general picture concerning modern soils and exotics in the regions concerned:

- Acrisols, having in certain depths low-activity clays and a low base status, occur in humid tropical, humid subtropical, and warm temperate regions; they are likely palaeosols in the present Mediterranean pedoenvironment (Verheye and de la Rosa 2005).
- Albeluvisols are not frequent at these latitudes.
- Alisols, having a low base saturation at certain depths, high-activity clays throughout the argic horizon, and toxic levels of Al at shallow depth, occur in humid tropical, humid subtropical, and monsoon climates (in places in temperate regions they are relict soils).

- Andosols are azonal soils but have a clear lithogenic mark, since they occur on pyroclastic materials in volcanic regions and all over the world. Their most important concentrations are on many islands in the Pacific and around its rim (along the so-called "ring of fire"). Nevertheless, it is by no means negligible that Andosols extend across the centre and east of the Mediterranean because of their usefulness in agriculture. Their susceptibility to erosion needs careful practices for soil conservation.
- Anthrosols clearly extend over wherever agriculture has been practised for a long time.
- Arenosols, being typical of shifting sands and active dunes, extend mainly from the fringe of the Sahara desert towards the south, although they also occur on sandy coastal plains and in coastal dunes in temperate regions.
- The Calcisols group accommodates soils in which there is substantial secondary accumulation of lime. They are common in highly calcareous parent materials and widespread in arid and semi-arid regions, mainly of the southern and eastern Mediterranean countries, where they represent an important productive resource in the absence of better soils. Petrocalcic horizons (calcrête) needs careful agricultural management (Fig. 4.2).

Fig. 4.2 Pistachio grove on Luvic Petric Calcisol (Chromic), with a petrocalcic horizon (calcrête) near the surface, probably formed on Saharan dust. A suitable soil management enabled the cultivation of typical crops in spite of the obvious limiting factors (Adana, Turkey). *Source* Photograph by courtesy of S. Kapur

- Cambisols are characterized by slight or moderate weathering of parent material and by the absence of appreciable quantities of illuviated clay, organic matter, and Al and/or Fe compounds. Cambisols with high base saturation in the temperate zone are among the most productive soils on earth. The most acid ones, although less fertile, are used for mixed arable farming and as grazing and forest land. Cambisols on steep slopes are best kept under forest. Rhodic and Chromic sub-types could partially correspond to the outdated Red Mediterranean soils or "Terra rossa" (Figs. 4.2 and 4.3).
- Chernozems can be found as relict soils, a witness to ancient steppe climates.
- Cryosols, which form in a permafrost environment, are obviously found only at the highest altitudes on mountain ranges.
- Durisols, containing cemented secondary silica, are mainly developed on alluvial and colluvial deposits, and are associated with old surfaces in arid and semi-arid regions bordering the Mediterranean regions.
- Ferralsols, being almost exclusive to humid tropics, are absent.

Fig. 4.3 Olive grove over reworked Calcic Luvisols (Chromic). The terracing enables cropping even on steep slopes and the control of erosion processes (Palestine, West Bank). *Source* Photograph by the author

- Fluvisols are very common in the Mediterranean region, as well as on all continents and in all climate zones. They develop on alluvial plains, river fans, valleys, and tidal marshes and lacustrine and marine deposits. River landscapes and marine plains have historically been places where great civilizations have developed.
- Gleysols are typically azonal soils and occur in nearly all climates, from per-humid to arid, wherever the groundwater saturates the soil for long enough periods.
- Gypsisols, characterized by secondary accumulation of hydrated calcium sulphate, which is very soluble, are obviously found in the driest parts of the arid and desert climate zone, mainly in the eastern and southern parts of the Mediterranean basin.
- Histosols, which occurr extensively in boreal, arctic, and subarctic regions, are restricted in the study area to depressions and poorly drained swamps, marshlands, and basins with shallow groundwater.
- Kastanozems, typical soils of dry and continental climates with relatively cold winters and hot summers, certainly exist in some Mediterranean countries, but not along the sea coasts (Fig. 4.4).
- Leptosols are very shallow soils, extremely gravelly and/or stony, over hard rock or highly calcareous material. Leptosols are azonal soils with an incomplete solum and/or without clearly expressed morphological features (Driessen et al. 2001). They are typical soils of the mountain areas, where pedogenesis is hindered by the erosion processes. In spite of this, they have a resource potential for wet-season grazing and as forest land. The Lithic (less than 10 cm in depth), Calcaric (rich in free calcium carbonate), and Rendzic (having a mollic horizon that contains or immediately overlies calcaric materials containing 40 % or more calcium carbonate) sub-types are the most widespread in the Mediterranean mountains.
- Lixisols have an argic subsoil horizon, a high base saturation, and low-activity clays at certain depths. They are not common in the Mediterranean region, but rather in more southern regions of Africa, with a tropical, subtropical, or warm climate with a pronounced dry season, notably on old erosion or deposition surfaces. Many Lixisols are regarded as polygenetic soils with characteristics formed under a more humid climate in the past, as well as at present in temperate countries.
- Luvisols have high-activity clays throughout the argic horizon and a high base saturation at certain depths (Fig. 4.3). They are most common in flat or gently sloping land in cool temperate and warm regions around the Mediterranean with distinct dry and wet seasons. Since they represent a great agricultural resource for the Mediterranean countries (Fig. 4.5), there is an urgent need to take the necessary steps to protect them from numerous misuses (settlements, roads, industries, quarries, etc.). Chromic, Calcic, Vertic, and Rhodic sub-types are almost equivalent to the Red Mediterranean soils (Terra rossa) of the past (Yaalon 1997).

Fig. 4.4 A deep soil profile (around 200 cm) probably formed under a past steppe climate, a little more humid than the present semi-arid one. The surface horizon is a mollic epipedon (50–60 cm deep), having a rather high organic carbon content (Kasserine, Tunisia). *Source* Photograph by the author

- Nitisols are deep, well-drained, red, tropical soils. In spite of some current mentions of these soils in the Mediterranean region, they seem rather to be remnants of past climatic conditions, probably of the Pleistocene period, because they predominate under tropical rain forest or savannah, with a level to hilly topography.
- Phaeozems cannot be considered typical soils of the Mediterranean countries, despite their scattered presence in moderately humid areas of meadows under continental climatic influences and sufficient organic carbon accumulation.
- Planosols have been produced either by lithological stratification, or by pedogenetic destruction or removal of clay or both. So they are relatively coarse-textured and have a light-coloured surface horizon abruptly overlying finer textured subsoil that impedes downward percolation of water and causes temporarily reducing conditions. These soils are not common in the regions concerned.
- Plinthosols are soils that are Fe-, Al-, and Mn-rich and organic-carbon-poor, containing a mixing of kaolinite (and sometimes gibbsite) and quartz sand.

Fig. 4.5 A Calcic Luvisol, with an unexpected thickness (200 cm), covered by 65 cm of Holocene aeolian sand. Depth and horizonation of the buried soil attest more humid environmental conditions in the past, compared with the present semi-arid climate (Kasserine, Tunisia). *Source* Photograph by the author

When plinthite by atmospheric exposure changes irreversibly to a layer with hard nodules or a hardpan, it becomes a cemented petroplinthite. Such soils are usually associated with rainforest and the savannah zone, but can occasionally occur in the Mediterranean pedoenvironment, where they usually represent remnants of palaeo-environmental conditions that existed in the Pleistocene interglacials or in the Tertiary era.

- Podzols are more typical soils developed over acidic parent materials in humid temperate and boreal regions of the northern hemisphere, under heather and/or coniferous forest, but also in some places in the humid tropics under light forest.
- Regosols, considered "the taxonomic remnant group" by WRB (2006), are poorly developed soils on unconsolidated materials (fine sands and silt, moraine, marls), lacking any zonality character and consequently widespread also in the region under discussion.
- Solonchaks are soils that have a high concentration of soluble salts at some time in the year, notably in areas where ascending groundwater reaches the solum. They are confined to the arid and semi-arid climate zones and to coastal regions

around the Mediterranean basin, mainly in North Africa and the Near East, with vegetation of grasses and/or halophytic herbs. The needs of arable lands in these countries stress the importance of appropriate irrigation techniques accompanied by drainage systems.

- Solonetz are soils with a dense, strongly structured, clayey subsurface horizon that has a high proportion of adsorbed Na and/or Mg ions. They normally occur on flat lands in climates with hot and dry summers. Major concentrations of Solonetz are over loess, loam, or clay in semi-arid, temperate, and subtropical regions.
- Stagnosols are soils with a perched water table showing redoximorphic features caused by surface water stagnating above a dense subsoil.
- Umbrisols are typical soils of humid climates in mountainous regions with little or no moisture deficit, in mostly cool areas but including subtropical mountains. They have a dark brown umbric surface horizon, and a low base saturation. The predominance of sloping land and wet and cool climate conditions restricts utilization of many Umbrisols to extensive grazing and forestry, due to their susceptibility to erosion. The planting of perennial crops and bench or contour terracing offer possibilities for permanent agriculture on gentler slopes.
- Vertisols are not exclusive to the Mediterranean environment, as they occur in tropical, subtropical, and semi-arid to subhumid and humid climates with an alternation of distinct wet and dry seasons. They form on sediments or weathered rocks that contain a high proportion of swelling clays. Alternate swelling and shrinking of expanding clays result in deep cracks during the dry season, and formation of slickensides and wedge-shaped structural elements in the subsurface soil. Vertisols are typically found in lower landscape positions such as dry lake bottoms, river basins, lower river terraces, and other lowlands that are periodically wet in their natural state. Having good chemical fertility and occurring on extensive level plains, they have considerable agricultural potential, from rain-fed up to irrigated crops, but adapted water management is a precondition for sustained production.
- Technosols, which have as parent material all kinds of materials made or exposed by human activity, are obviously widespread everywhere industrial activities, buildings, mines, quarries, and dumps have covered or exposed originally buried raw materials. In spite of the lack of detailed maps of these technogenic soils in the Mediterranean countries, they cover vast and increasing areas, not only in industrialized countries but also in the developing ones.

On the whole, according to Verheye and de la Rosa (2005), we can consider that, of the roughly 420 Mha occupied by Mediterranean soils, the prevailing soils are Calcisols (85 Mha), followed by Cambisols (68 Mha), and Luvisols (65 Mha), not counting Leptosols.

Particular investigations, classification, and mapping should require the *palaeosols*, which at present, in spite of several attempts and advanced proposals (Catt 1998; Nettleton et al. 1998, 2000) do not yet have a taxonomy shared and applied at the international level. The palaeosols of the Mediterranean regions, in their variants buried, exhumed, and relict (Nettleton et al. 2000), are mainly widespread as spots in areas not covered by ice during the Pleistocene or spared by the glacial erosion.

4.4 Soil Direct and Indirect Threats

Soil consumption, misuse, and degradation eventually leading to desertification dates back a long time in the Mediterranean regions and has different causes (Madrau et al. 2010): worsening climate, wars, fires, logging, grazing, urbanization, and so on. Many civilizations fell into decline and became extinct through dramatic forest clearance, together with soil over-exploitation, erosion, and salinization.

House building and soil sealing, frequently carried out on the best agricultural soils, a process increasing today, have always been the main threats to such a precious productive and environmental resource (Tóth 2002). Population pressure, water shortages, and the risk of desertification also threaten many Mediterranean countries, and together with the development model adopted in many countries could lead the whole area into an irreversible crisis. According to Zdruli et al. (2008, 2010), arable land per capita in 2020 could be 0.22 ha/person compared to 0.48 ha/person in 1961. In 2020 the region will have in total more than 500 million people, while in 2000 the population was around 400 million.

4.4.1 Erosion

The aggressiveness of the Mediterranean climate is mainly connected with the rainfall pattern. Most rain (around 90 % of the total annual) falls in autumn and winter, with intensity per hour particularly high, reaching 100 mm/hr. Steepness is a natural predisposing factor to erosion, but slope exposition, rock type and its spatial disposition, soil depth, texture, organic matter content, and porosity, together with vegetation cover and management practices, all play an important role in the process. More than 30 % of the losses of Mediterranean land are caused by erosion, at a rate of more than 15 t/ha. Experts estimate that in the EU, 26 million hectares are subject to water erosion and 1 million to aeolian erosion.

The run-off and the consequent soil erosion are more intensive where over-grazing, downhill and deep tillage, wildfires, new roads, quarries, and other anthropic impacts have aggravated the fragile equilibrium between soils, water, and vegetation.

In addition, the eastern and southern arid and semi-arid regions suffer from aeolian erosion that abrades soils while in many areas accumulating sandy sediments. All these processes of erosion are worsened by the weak stability of soils impoverished in organic matter content. At least 75 % of the world's farmed soils have been degraded or affected by erosion.

The greatest amount of erosion is occurring now on agricultural and afforested land. When erosion occurs, it can also cause flooding and off-site damage on flood plains. Several different types of erosion occur in the Mediterranean region and these include splash and sheet erosion, rill, gully, tunnel and channel erosion, and wind erosion, as well as erosion caused by animals and land use activities. The problems are potentially worse than in more humid regions because of the specific soil and climate conditions.

The principles of soil conservation and protection have probably been understood for millennia, so that when erosion problems occur in the Mediterranean, it is a simple matter to take appropriate action to stop it. Soil conservation techniques, accompanied by careful consideration of soil capability and suitability, today enables the annual soil losses to be successfully minimized and prevented (Zucca et al. 2011a, b).

4.4.2 Organic Matter Decline

A sufficiently high level of organic matter is important for increasing soil stability, root penetration, the water-holding capacity, and the nutrient holding capacity and supply. Organic matter enhances the soil's resistance to erosion and guarantees its fertility, contributing through its binding and buffer capacity to limiting the diffusion of soil and water pollution. The soil organic matter (vegetal remains, living organisms, humus) releases carbon into the atmosphere in the form of CO_2, which is then sequestered in the process of photosynthesis.

Many Mediterranean soils—mainly Leptosols, Calcisols, Regosols, Arenosols, and some Cambisols—suffer from a natural lack of organic matter. An organic matter content of less than 1.7 % (1 % C_{org}) is usually considered low and indicates a risk of degradation and/or desertification. This constraint damages current productivity and compromises the future ability of the soil to yield enough. Therefore, additional organic and/or inorganic fertilization is necessary to restore and maintain the optimal productivity of the soil. Some of the usual measures for maintenance of organic matter are: minimum-tillage and no-tillage, biological agriculture, permanent pastures, cover crops, mulching, green manure, stable manure and compost, strip cultivation, and contour farming.

4.4.3 Salinization, Alkalization, and Acidification

Two kinds of salinity are known: the primary salinity of soils that derives from the proximity of the sea or from salt release from geological deposits, and secondary salinity, related mainly to anthropogenic activities, especially improper irrigation without drainage systems or the use of salt-rich waters. The rise in level of the saline water table because of over-abundant irrigation or poor drainage is also a frequent reason for soil salinization, as is the intrusion of sea water into freshwater aquifers. The latter is a frequent problem in coastal areas, due to the over-exploitation of water resources.

The natural saline soils (Solonchaks) are frequent in arid and semi-arid regions where the potential evaporation of the soil greatly exceeds the quantity of water that infiltrates the soil. Salt modifies soil properties, plant growth, and soil erosion, and consequently could trigger desertification. In southern Europe, scattered areas of intensive agriculture are subjected to salinization (Zucca et al. 2011a).

Salinity has major economic consequences because it degrades the fertility of cultivated areas and can prevent or strongly limit the agricultural use of the land affected. The soluble salts in the soil have a depressive effect on plant growth after a threshold that varies according to the different species.

It is calculated that there are more than ten million ha of saline soils in the Mediterranean regions which are seasonally or permanently involved in areas of high evaporation or the intrusion of saline groundwater (Varallyay 2006).

A worse process that occurs in some soils is the alkalization that corresponds to the adsorption on clay particles of sodium instead of potassium and calcium. Alkalinity can be either primary and natural or secondary and induced by anthropic activities. The process leads to the rapid degradation of the soil structure through the dispersion of colloidal inorganic and organic substances. The permeability of alkaline soil (Solonetz) is generally very low, thus reducing water and air capacity, and the pH values are higher than 8.0. Consequently, most plants do not tolerate such conditions, except for a few adapted species. This process occurs particularly on desalinated soils by the sea or, more recently, on farming land that has suffered a heavy spillage of sodium-rich farming effluents (Middleton and Thomas 1997).

A third chemical process affecting soils is acidification, induced by several causes:

- the importation of certain acidifying fertilizers by agriculture;
- dry or acid-rain deposition of pollutants (e.g. sulphur dioxide, nitrogen oxides) emitted by vehicles, boilers, thermoelectric power plants and certain industries;
- the massive planting of certain softwood trees (spruce) and certain hardwood trees (eucalyptus).

Acidification can lead to soil depletion of minerals such as magnesium, potassium, and calcium, or to the mutation of other minerals such as aluminium and manganese into a form which is toxic for plants. Downstream of acidified watersheds, streams can be acidified, and this disturbs the fauna and flora (de Franchis 2003).

4.4.4 Landslides

There is a relationship between landslides and soil properties. Jones (2006) has demonstrated that areas with organic carbon (OC) of less than 1.0 % (1.7 % organic matter) suffer a soil loss of more than 2 t/ha/year. But we must consider that the predisposing factors for slides, falls, and debris flows are more complex and numerous: height and gradient of the slope, aspect, rainfall distribution, relative proportion of water to solid material, lithology, soil type, strata inclination, vegetation cover, and so on. Certainly, all the latter conditions being equal, the organic carbon content becomes crucial, because of the role it plays in relation to soil cohesiveness and erodibility. Shallow and fragile soils, such as Leptosols, Andosols, Regosols, and Arenosols are more prone to all the different forms of erosion. In addition, Vertisols favour the triggering of mudslides and earthflows. Numerous agricultural practices, land uses, and connected effects, such as deep tillage on steep slopes, land levelling, harvesting of root crops, animal trampling, overgrazing, and heavy mechanization, are frequent causes of sheet, linear, and mass erosion.

In addition, the construction of roads in landscape-sensitive areas can increase the frequency of landslides and amplify the material damage caused, sometimes with tragic consequences. The development of uncontrolled constructions on unstable land located on steep hills near cities can lead to very serious disasters when these lands collapse in heavy rains or earthquakes.

4.4.5 Urbanization and Soil Sealing

"Artificialization" designates the phenomenon of expanding built-up areas to the detriment of natural or agricultural soils. This means urbanized areas, industrial or commercial zones, communication networks, mines, landfill sites, and building sites. It leads to the irreversible disappearance of an increasing part of the soil resource. It affects all Mediterranean areas, in particular former agricultural areas located on the outskirts of cities. The impact on landscapes and the organization of space is different, depending on the regions and countries, and most especially on the type of artificialization. For example, urban sprawl will have very different consequences depending on whether it is organized according to a plan or if it occurs in an uncontrolled, even illegal, fashion. Land artificialization implies more or less extensive sealing of the soil, which restricts the infiltration of water into the ground and the replenishment of the water table, the whole increasing the quantity of water to be evacuated by run-off.

Urbanization and "littoralization" (Zdruli et al. 2008, 2010) are connected processes typical of the Mediterranean countries. Generally, littoralization could be described as the migration of the population of the interior towards the coast and the "maritimization" of the economy linked with productive activities, all resulting in a rapid and intense expansion of artificial land cover. Large inland areas are being abandoned, leaving neglected forests, eroded lands, collapsed terraces, overgrazing, forest fires, and overall degradation. The most fertile soils, often located downstream on plains, are threatened by large-scale urbanization and littoralization induced by demographic growth, country-to-town population migration, and the evolution of lifestyles and production systems.

Fertile soil sealing (arable land, wetlands) by urbanization represents nearly irreversible degradation and would imply serious, long-term consequences if the present trends were to be confirmed (Tóth 2002).

4.4.6 Impacts of Agriculture

Agriculture is not always an activity that protects the environment. Modern farming, the systematic use of chemical fertilizers and pesticides, and the irrational development of irrigation can lead to the accumulation of strong concentrations of toxic substances in soils. Nitrates and phosphates are not strictly pollutant but they become so in excess, when they are leached and lost from the soil. Furthermore, biocides (pesticides, fungicides, herbicides) have a negative impact on soil fauna and flora, and reduce the organic matter content of the soils. This pollution can be apparent on-site or downstream and off-site following transport of the products. Nitrates are transported as dissolved salts by run-off or infiltration water. Phosphates and pesticides are transported as molecules or crystals bonded to soil particles carried off by erosion.

Agricultural additions of pesticides, nitrates, and phosphates change the soil's biochemical balances and thus its properties. The positive effects of these modifications (increase in fertility, reduction of diseases) are obviously the effects aimed for by farmers. Nitrates heavily modify the nitrogen cycle (especially the C/N ratio), the degree of decomposition of organic matter, and then the structure of soils and their water storage capacity. In cases of poorly managed fertilization there can be local soil degradation (e.g. acidification, slaking crust). Water-carried nitrates pollute streams and groundwater. They can cause more or less serious eutrophication of rivers, standing water, and coastal areas, leading to the degradation of biodiversity, fishery resources, and the landscape. They cause additional costs for water treatment (EEA/UNEP 1999) and can even make water unfit for consumption.

Phosphates are also a major cause of eutrophication. Once erosion has deposited phosphor-rich sediment at the bottom of natural or human-made standing water, it is very hard to eliminate it. Phosphates are trapped in these sediments, which then constitute a permanent threat of water eutrophication.

4.4.7 Contamination by Mines, Industries, Waste Disposals, and Dumps

Since the beginning of industrial civilization in the middle of the eighteenth century, sources of pollution created by humankind have increased in intensity, number, and extent (EEA/UNEP 1999). Among the kinds of soil pollution most often mentioned in the Mediterranean area are salinization and heavy metal pollution, and the data on soil pollution in these regions show that the total areas concerned are relatively limited compared to those affected by erosion or anthropogenic occupation. But the danger from heavy metals is high in spite of their localization.

High concentrations of heavy metal can be either natural (due to mineralogical composition of the geological formations) or human-induced, due to direct or indirect input related to human activities. The most dangerous elements for mankind and animals are mercury, lead, cadmium, and arsenic. Copper, nickel, and cobalt are only dangerous in high concentrations. Soils naturally contain a lot of chemical elements, in particular metals, but the level of danger depends not only on concentration, but also on the chemical forms in which they present (total, free, bonded with organic matter, etc.). Heavy metal input created by humankind is often present in a more highly reactive form and thus more dangerous than naturally occurring minerals. Besides this, buffer soil capacity, plant cover, and climate conditions greatly influence the evolution and fate of the metals.

It is not only factories and active or abandoned mines, but also the dumping of agricultural, industrial, and household waste that constitute a dangerous source of metallic pollutants. The spreading of phosphate fertilizers (containing cadmium), pork slurry (with copper and zinc), and sludge from purification plants constitutes, together with industrial pollution and car traffic, the fundamental source of soil contaminants. Furthermore, the manifold pathways of pollution are represented by the use and consumption of industrial products.

The restoration costs of soils polluted by heavy metals are almost always extremely high. The techniques used range from excavating the soil and eliminating the waste off-site to covering the ground with an impermeable material to avoid contact with water and to reduce the risk of pollutants leaching towards water tables. Purifying water usually means pumping and treating it. More advanced technologies such as treating contaminated sites on-site are at present being tested.

As with all problems of soil degradation, prevention is always less costly than the rehabilitation of polluted soils.

4.4.8 Deforestation and Fires

Dating back some millennia (Fig. 4.6), the practice of deforestation has caused not only the stripping of vegetation cover, but frequently also the removal of soil cover, together with the loss of huge quantities of CO_2. The process is frequently accompanied by or caused by forest fires and has assumed a planetary dimension and relevance. A large area of tropical forests and mangroves is destroyed daily, with great damage to soils, waters, lakes, sea, biodiversity, and air quality. The semi-arid countries of the Mediterranean region have undergone severe deforestation during the past centuries, with serious consequences for the way of life of the local people. It is clear that the political and economic aspects of this critical world problem need decisions at the international level.

Focusing solely on the technical aspect of the problem, it is evident that reforestation is closely connected with interventions that attempt to control the processes of erosion. An eroding slope will eventually fail if regressive erosion continues to deepen the bed of the waters that run at the foot of the slope. Conversely, in some cases there can be spontaneous natural reforesting of slopes once the deepening of the erosion has been stopped.

Fig. 4.6 Typical landscape of central Morocco. The effects of historical deforestation are evident. In the background can be seen the Jebilet massif. (Rural Municipality of Ouled Dlim; 30 km north-west of Marrakech.) ***Source*** Photograph by the author

Fire is a natural factor in most Mediterranean ecosystems; however, the increased frequency and extent of wildfires since the mid-twentieth century have made wildfires one of the most important driving forces for desertification under dry Mediterranean conditions. Today it has been proved that the majority of fires are deliberate or due to carelessness. Many mineralogical, physical, biological, and chemical soil properties can be modified by forest fires, but their effects can be different depending on temperature and duration of burn. The resilience of the pedoenvironment depends on the climate, vegetation, and topography of the area concerned. Fires of moderate intensity, such as managed ones, eliminate weeds and increase pH and the availability of nutrients (Certini 2005), but at the same time can enhance the hydrophobicity and erodibility of the soil. Long-lasting wildfires usually cause significant removal of organic matter, deterioration of both structure and porosity, considerable loss of nutrients through volatilization, ash entrapment in smoke columns, leaching and erosion, and marked alteration of both the quantity and the specific composition of microbial and soil-dwelling invertebrate communities (Certini 2005). Of course, if early rain stimulates the process, the recolonization of burnt forest can be rapid and successful.

The assessment of the vulnerability of ecosystems to wildfire is fundamental to the planning of post-fire management, and it is estimated on the basis of soil erodibility, climate, topography, and the capacity of vegetation for recovery.

The accurate identification of areas sensitive to degradation after fire and the implementation of suitable restoration techniques will help mitigate the impacts of fire and enhance post-fire regeneration towards mature, less flammable forest ecosystems. In any case, the multiplication of dwellings in wooded areas increases the number of accidental fires and raises the cost of fighting them. In some countries, the rise in living standards has led to an increase in the number of private homes built in pine groves, and at the same time an increase in the number of forest fires (Conacher and Sala 1998).

4.4.9 Destruction and Contamination by Bombing

A particular and usually unmentioned form of "anthropic erosion", instantly devastating and deliberately caused, is represented by the effects of bombing, gunfire, and explosions connected with actions of war. Such activities, besides destroying human lives, animals, plants, infrastructure, and dwellings, also contaminate the air and water and disrupt soil, and the environmental consequences frequently remain for a long time (Reuveny et al. 2010). A large amount of toxic metals are unloaded into the environment, in particular tungsten, mercury, molybdenum, cadmium, and cobalt. Excessive amounts of these metals can cause tumours and problems with fertility, and they can have serious effects on newly-born babies, such as deformities and genetic pathologies (Austin et al. 2000). Even the topography and landforms are deeply modified by military actions, forcing the revision of the topographic maps of the lands concerned, as has happened in

Lebanon, Palestine, Kosovo, Serbia, and Libya not to mention Afghanistan, Iraq, and Syria.

Furthermore, in the past fifty years not only has environmental pollution been intensified by radioactive fallout from nuclear bomb testing, spillage from nuclear power plants, and dumping of waste, but the many local wars have been typified by the use of new arms and toxic substances in defiance of the Geneva Convention, such as depleted uranium and white phosphorus. These substances, if trapped in a biological replicating system, may cause genomic amplification of damage over time (Busby 2005). Such uses are similar to the use of chemical or biological weapons banned by the Geneva Convention. The embarrassing question is whether mankind really needs to use these instruments just to resolve international disputes.

4.4.10 Sport Resorts in Mountains

The mountain regions around the Mediterranean suffer damage not only from the above-mentioned threats but also from the less-investigated impact of resort development in many mountain localities; some of these types of damage are not much known. This kind of land use has multiple and serious impacts on an environment as sensitive as the mountains (de Jong and Barth 2008; de Jong et al. 2009; Previtali 2011). The main impacts are:

- land impermeabilization through urbanization and soil sealing by ski runs;
- vegetation removal and excavation and handling of earth to create lift tracks, roads, and water reservoirs for snow production, which all accelerate erosion and trigger landslides and debris flows;
- input of contaminating elements and organisms (hydrocarbons, heavy metals, wastes, bacteria and chemicals for artificial snow, etc.);
- water shortage and conflicts (drinking water, artificial snow production, swimming pools, and household consumption);
- alteration of the superficial hydrography and the hydrogeological arrangement.

It should be noted that in the Mediterranean countries ski resorts have expanded not only in the well-known areas of Spain, France, Italy, Slovenia, Croatia, and Bosnia and Herzegovina, but also in Serbia, Greece, Turkey, Lebanon, Syria, Israel, Tunisia, Algeria, and Morocco. In the countries of both the first and second groups, the decreasing depth of the snow mantle and the shortening of its time on the ground are forcing a growing recourse to artificial snow and the raising to higher altitudes of lifts and ski runs.

It is pointless to stress that in some of the above-mentioned countries environmental conditions are inconsistent with the development of winter tourism mainly based on downhill skiing, with a great impact on the mountain environment. Everywhere, the establishment and development of these resorts is joined with deforestation, erosion, uncontrolled building, disruption of the landscape and natural habitat, air, water, and soil pollution, and water shortage. As de Jong et al. (2009) effectively argue, all the reasons expounded make this kind of exploitation of mountain resources unsustainable, from both an economic and an environmental point of view.

4.5 Conclusions

Agriculture has a major economic importance in most of the Mediterranean countries, and in the southern and eastern ones represents over 15 % of the gross domestic product (de Franchis 2003). It thus constitutes a basic source of employment, especially in regions where it still remains the main activity.

Conservation of the Mediterranean agricultural potential therefore constitutes a challenge for maintaining the social and economic structure and the food security of societies.

The degradation of lands can lead to their abandonment and to the flow of rural populations to the cities, which poses serious economic and social problems in terms of regional planning and employment. So, it is necessary to maintain rural spaces capable of sustaining a large population in satisfactory economic and social conditions (EEA/UNEP 2000; Zalidis et al. 2002; Verheye and de la Rosa 2005; Zdruli et al. 2008; Zucca et al. 2011a, b).

Well-managed agriculture plays a fundamental role in the management of natural resources through water harvesting systems, terracing, controlled pastoralism, tree plantations, and so on. For instance, the abandonment of management systems such as the old terraces is leading in some regions to the irreversible degradation of slopes.

Soils are basic components of ecosystems and their degradation generally has a strong impact on all other segments, such as flora, fauna, and water and nutrient cycles. Wooded areas, oases, wetlands, and steppes, which constitute a precious heritage of the Mediterranean landscape, are under threat due to irrational clearing, pre-desertification, contamination, agricultural over-exploitation, overgrazing, urbanization, and so on.

There is an urgent need for careful planning of both urban development and farming where ecological aspects are included in the technical procedures.

Acknowledgments The author is indebted to Prof. Dr. Selim Kapur for his useful and relevant advice and remarks.

References

Austin, J.E., Bruch, C., (eds.): 2000: "The Environmental Consequences of War: Legal, Economic, and Scientific Perspectives", (Cambridge, UK: Cambridge University Press).

Benoit G., Comeau A., 2005: "A Sustainable Future for the Mediterranean", in: *The Blue Plan's Environment and Development Outlook*. (London: Earthscan).

Busby C., 2005: "Depleted Uranium weapons, metal particles and radiation dose", in: *European Journal Biology and Bioelectronics,* 1, 1 82–93.
Catt J. A., 1998: "Report from Working Group on Definitions Used in Paleopedology", in: *Quaternary International*, 51/52, 81–86.
Certini G., 2005: "Effects of Fire on Properties of Forest Soils: a Review", in: *Oecologia*, 143, 1–10.
CIHEAM 1993: Les sols dans la région méditerranéenne: utilisation, gestion et perspectives d'évolution. Institut Agronomique Méditerranéen de Zaragozza, 1993). CCE-DG I, Cahiers Options Méditerranéennes, vol. 1, n 2. Zaragoza. 269 p.
Conacher A.J., Sala M., 1998: "Land Degradation in Mediterranean Environments of The World: Nature and Extent, Causes and Solutions", (Chichester: Wiley).
de Franchis L., 2003: "Threats to Soils in Mediterranean Countries", in: *Plan Bleu-Centre d'activités régionales*. Sophia Antipolis.
de Jong C., Barth T., 2008: "Challenges in Hydrology of Mountain Ski Resorts under Changing Climatic and Human Pressures", in: *ESA Proceedings*, Geneve.
de Jong C., Lawler D., Essery R., 2009: "Mountain Hydroclimatology and Snow Seasonality—Perspectives on Climate Impacts, Snow Seasonality and Hydrological Change in Mountain Environments", in: *Hydrol. Process,* Published online in Wiley InterScience.
Driessen P., Deckers J., Nachtergaele F., 2001: "Lecture Notes on the Major Soils of the World", *World Soil Resources Reports* 94. FAO, Rome.
EEA (European Environment Agency)/UNEP 1999: "State and pressures of the marine and coastal Mediterranean environment", (Luxembourg: Office for Official Publications of the European Communities): 137 p.
EEA (European Environment Agency)/UNEP 2000: "Down to earth: soil degradation and sustainable development in Europe", *A challenge for the* 21*st century,* (Luxembourg: Office for Official Publications of the European Communities): 32 p.
FAO 1976: "A framework for land evaluation", *FAO soils bulletin,* 32, Rome.
IUSS Working Group WRB 2006: "World reference base for soil resources 2006", 2nd edition, in: *World Soil Resources Reports* No. 103. FAO, Rome.
Jones B., 2006: "Soil Erosion, Organic Matter Decline and Landslides", in: *4th European Summer School on Soil Survey*, JRC, National Soil Resources Institute Cranfield University. 28 Aug–1 Sept, Ispra (I).
Madrau S., Zucca C., Urgeghe A.M., Julitta F., Previtali F., 2010: "Land Suitability for Crop Options Evaluation in Areas Affected by Desertification: The Case Study of Feriana in Tunisia", in: Zdruli, P., Pagliai, M., Kapur, S., Faz Cano, A., (eds.): *Land Degradation and Desertification: Assessment, Mitigation and Remediation*, First Edition, Springer Science.
Middleton, N., Thomas, D., (eds.): 1997: "World Atlas of Desertification", (London: Arnold): 182 p.
Morgan R.P.C., 2005: "Soil erosion and Conservation", Third edition, (Oxford, UK: Blackwell publishing): 303 p.
Nettleton W.D., Brasher B.R., Benham E.C., Ahrens R.J., 1998: "A Classification System for Buried Paleosols", in: *Quaternary International*, 51/52, 175–183.
Nettleton W.D., Olson C.G., Wysocki D.A., 2000: "Paleosol Classification: Problems and Solutions", in: *Catena*, 41, 61–92.
Previtali F., 2011: "Mountain Anthroscapes, the Case of the Italian Alps", in: Kapur, S., Eswaran, H., Blum, W.E.H., (eds.): *Sustainable Land Management—Learning from the Past for the Future*. (Heidelberg, Springer-Verlag).
Reuveny R., Mihalache-O'Keef A.S., Li Q., 2010: "The Effect of Warfare on the Environment", in: *Journal of Peace Research,* 47, 6 749–761.
Soil Survey Staff 1999: "Soil Taxonomy: A Basic System of Soil Classification for Making and Interpreting Soil Surveys", 2nd Edition. in: *USDA*, Agriculture Handbook No. 436. Washington DC.

Tóth G., 2002: "Soil Functions and Soil Sealing", in: *European Commission Joint Research Centre*, Institute for Environment and Sustainability, Land Management and Natural Hazards Unit.
Van Lynden G.W.J., 2000: "Guidelines for the Assessment of Soil Degradation in Central and Eastern Europe", *Report* 97/08b. FAO&ISRIC.
Varallyay G., 2006: "Salinisation/Sodification", in: 4th *JRC Int.School on Soil Survey*, Ispra (Italy), 28 Aug.
Verheye W., de la Rosa D., 2005: "Mediterranean Soils". in: *Land Use and Land Cover*. Encyclopedia of Life Support Systems (EOLSS), Developed under the Auspices of the UNESCO, (Oxford, UK: Eolss Publishers).
Yaalon D.H., 1997: "Soils in the Mediterranean region: what makes them different?", in: *Catena*, 28, 157–169.
Zalidis G., Stamatiadis S., Takavakoglou V., Eskridge K., Misopolinos N., 2002: "Impacts of agricultural practices on soil and water quality in the Mediterranean region and proposed assessment methodology", in: *Agriculture, Ecosystems and Environment*, 88, 137–146.
Zdruli P., Lacirignola C., Trisorio Liuzzi G., 2008: "Land degradation in the Mediterranean: Findings of the EU-funded MEDCOASTLAND Project", in: 5th *International Conference on Land Degradation*, Valenzano (Bari, Italy), 18–22 Sept.
Zdruli, P., Pagliai, M., Kapur, S., Faz Cano, A., (eds.): 2010: "Land Degradation and Desertification: Assessment, Mitigation and Remediation", Springer 490 p.
Zucca C., Julitta F., Previtali F., 2011a: "Land restoration by fodder shrubs in a semi-arid and agro-pastoral area of Morocco", in: *Effects on soils*. Catena, 87, 306–312.
Zucca C., Previtali F., Madrau S., Akça E., Kapur S., 2011b: "An Anthroscape from Morocco: Degraded Rangeland Systems and Introduction of Exotic Plant Material and Technology", in: Kapur, S., Eswaran, H., Blum, W.E.H., (eds.): *Sustainable Land Management—Learning from the Past for the Future*. (Heidelberg: Springer-Verlag).

Chapter 5
Carbon Cycle and Sequestration in Terrestrial Ecosystems with Specific Reference to Mediterranean and Boreal Regions

A. R. Mermut

Abstract Soil functions as a sink and contributes to the process of CO_2 reduction in the atmosphere. Benefits of terrestrial sequestration of carbon are well understood. About 75% of terrestrial carbon occurs in the soil and therefore soils are essential for carbon sequestration. The maximum capacity for sequestering carbon has not been well established. There is a strong need for applied research to determine the actual values that can be used to calculate the economic benefits of carbon sequestration. Numerous methods of sequestering carbon in agricultural land have been tested. Understanding how to increase soil carbon stocks in agricultural lands is critical to increasing the sustainability of food production. The lands in the Mediterranean area are characterized by a wide diversity of soils, landscape, vegetation, and formation time and, especially, the long-term influence of human activities. The effect of anthropogenic intervention on *soil organic carbon* (SOC) in these ecosystems is unknown. Strategies for sustainable soil management in Mediterranean regions include conserving soil organic matter, minimizing erosion, enhancing soil fertility, and balancing production with environmental sustainability, especially in areas subject to drought. Increasing numbers of research studies provide an excellent insight into understanding soil organic matter stocks and possibility sequestering carbon in soils around the Mediterranean region. Prairie soils in Boreal Canada have likely lost more than 50% of their original carbon. However, studies have suggested that with conservation tillage and adequate fertilization, a significant increase in soil organic carbon over present-day levels could be sequestered with time and consequently increase soil quality and productivity.

Keywords Terrestrial ecosystem · Carbon sequestration · Carbon cycle · Mediterranean region

---

A. R. Mermut (✉)
Harran University Şanlıurfa, Turkey and University of Saskatchewan, Saskatoon, Canada
e-mail: a.mermut@sask.ca

S. Kapur and S. Erşahin (eds.), *Soil Security for Ecosystem Management*, SpringerBriefs in Environment, Security, Development and Peace 8,
DOI: 10.1007/978-3-319-00699-4_5,

5.1 Introduction

In the past decade, increasing awareness of CO_2 build-up in the atmosphere and the threat of global warming has encouraged all concerned societies interested in global warming to find means of reducing atmospheric CO_2. The concept of greenhouse gas reduction by sequestering carbon in different terrestrial ecosystems, or withdrawal of CO_2 from the atmosphere, has been extensively discussed over the last two decades. It is generally agreed that carbon sequestration can be a highly cost-effective and environmentally sound mitigation technique.

Despite the socio-economic and environmental benefits, societies interested in the area of carbon flux have not yet succeeded in translating the available knowledge of carbon dynamics into real agronomic practices. There is a great interest in carrying out practical work to design appropriate strategies for carbon sequestration. The research so far carried out on the terrestrial carbon cycle has confirmed the many agricultural and environmental benefits. Increased funding for research and development to address the practical implementation of carbon sequestration is of paramount importance.

There is general agreement that with appropriate management technologies soil can function as a sink and contribute to the process of CO_2 reduction in the atmosphere. This means drawing CO_2 out of the air and converting it to biomass (plants) or to *soil organic matter* (SOM). By using water and energy from the sun, plants are naturally capable of converting CO_2 to carbohydrates or biomass and consequently to organic matter in the soil.

The many benefits of terrestrial sequestration of carbon are also well documented (US Department of Energy 1999). These are:

- improving soil and water quality;
- decreasing plant nutrient loss;
- reducing soil erosion and increasing water conservation;
- providing better wildlife habitats and more biodiversity;
- creating conditions for higher plant productivity and more biomass products; and
- increased carbon sequestration will have the additional benefits of restoring degraded ecosystems worldwide.

Preliminary estimates suggest that, using appropriate management techniques, the ~40–80 Pg C that would be produced through combustion over the next fifty to a hundred years could be sequestered in cropland. This would mean that carbon sequestration offers a means of controlling the CO_2 levels in the air and keeping them below the critical threshold level of 550 ppm.

The rate of the carbon sequestration process is estimated to be ~2 Pg C/year (US Department of Energy 1999). About 75% of terrestrial carbon occurs in the soil and therefore soils are essential in terms of carbon sequestration. The potential for carbon sequestration appears to be great (5–10 Pg C/year) in comparison with the current rate for terrestrial ecosystems, if all terrestrial ecosystems are considered. The maximum capacity for sequestering carbon has not been well established. Table 5.1 shows estimates of the carbon sequestration potential of some major land use types.

5.2 Approaches to Sequestering Carbon in the Soil

There are two fundamental approaches to sequestering carbon:

1. the protection of ecosystems that store carbon so that sequestration can be maintained (increased residence time), and
2. the manipulation of ecosystems to increase carbon sequestration beyond the current conditions.

All other factors being equal, the rate of C sequestration in soils is higher for warm humid regions than for dry cool regions, and for severely degraded soils than for undegraded soils. It may seem unlikely that large amounts of carbon could be sequestered in dryland regions in comparison with other agro-ecological zones of the world. According to UNEP (1997), however, dryland stores sixty times more carbon than the carbon added to the atmosphere by fossil fuel. Dryland covers an area of 450 million ha throughout the world. A small change in the rate of carbon sequestered in dryland regions can have a large impact on CO_2 in the atmosphere. Over one billion people currently live in susceptible drylands, and any effort to restore the productivity of these eco-regions will be of benefit for their inhabitants.

The potential for carbon sequestration in dryland areas is shown in Fig. 5.1. While the drylands may be a source of atmospheric CO_2, if properly managed they can sequester 37 Pg C/year, which is about 15% of atmospheric CO_2 emissions (UNEP 1997). This would be a significant contribution to the mitigation of global warming.

Technological options so far recognized for sequestering carbon in agricultural land are few. Some have been developed for temperate and tropical regions and for developed and developing economies. Their application may differ from one region to another. Table 5.2 shows rather simple broad estimates for agricultural practices for different eco-regions. There is a strong need for applied research to determine the actual values that can be used to calculate the economic benefits of carbon sequestration.

One of the fundamental arguments is that more than 50% of SOM is lost in the topsoil through intensive agricultural practices. Uncultivated soils were in equilibrium with the native vegetation and accumulated large reserves of *soil organic carbon* (SOC) and cultivation has disrupted the steady-state equilibrium (Lal et al. 1999). There are reliable estimates that many cultivated soils in North America have lost substantial amounts of SOM in cultivated lands (Acton and Gregorich 1995; Bruce et al. 1999), and that this has resulted in a decline in production and an increase in soil erosion and soil degradation. The lost carbon should be urgently

Table 5.1 Estimates of carbon sequestration potential of some major land use types with projected annual carbon storage and time frames. *Source* UNEP (1997)

Option	Area million ha	Rate t C/ha^{-1}/year^{-1}	Period year	Cost US$/t C	Total Mt C/year^{-1}
Dryland crop management	450	0.3–1.0	5–20	1–5	135
Halophytes	130	0.5–5.0	Indefinite if harvested, 5 years if not	170 (irrigated and harvested) 20 (dryland not harvested)	65
Bush encroachment	150	0.1–0.5	15–50	10–20	37
Energy crops	20 (5% of dryland crop area)	4–8	Indefinite	2–5	80
Domestic biofuel efficiency	Not applicable	Not applicable	Indefinite	2–5	75
Agroforestry (arid)	50	0.2	30	2–10	10
Agroforestry (semi-arid)	75	0.5	20	2–10	38
Agroforestry (subhumid)	150	1.5	15	2–10	225
Improved pasture (semi-arid Asia)	10 (2500 degraded globally)	0.1	30	10	1
Savanna fire control	900 (globally)	0.5	30	1–5	450
Woodland management	400 (globally)	0.5	30	1–5	200

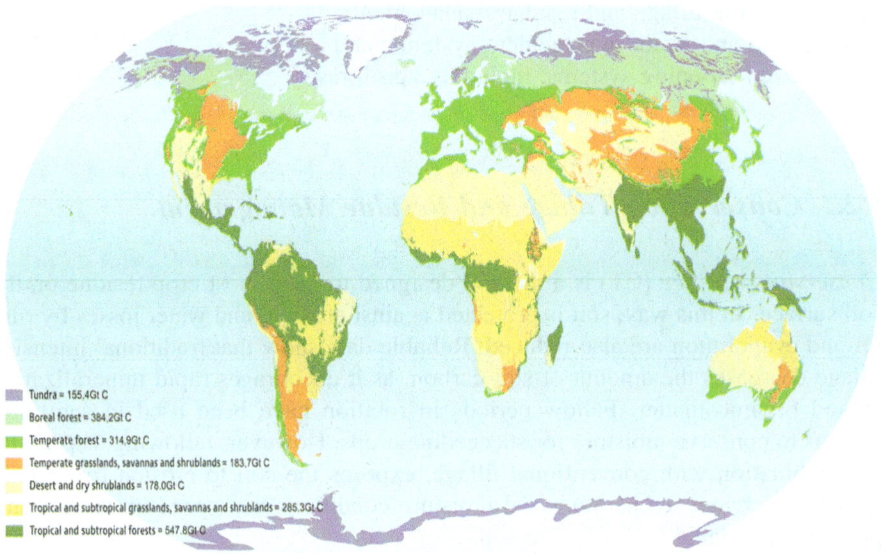

Fig. 5.1 Potential carbon sequestration by vegetation type. *Source* Kapos et al. (2008), Olson et al. (2001), UNEP (2009)

Table 5.2 Technological options for C sequestration in soil (t/ha/year). *Source* UNEP (1997)

Technological options	Temperate climate		Tropical and subtropical	
	Humid	Semi-arid	Humid	Semi-arid
1. Conservation tillage	0.5–1.0	0.2–0.5	0.2–0.5	0.1–0.2
2. Mulch farming(4–6 Mg/ha/year)	0.2–0.5	0.1–0.3	0.1–0.3	0.05–0.1
3. Compost (20 Mg/ha/year)	0.5–1.0	0.2–0.5	0.2–0.5	0.1–0.2
4. Elimination of bare fallow	0.2–0.4	0.1–0.2	0.1–0.2	0.05–0.1
5. Integrated nutrient management	0.2–0.4	0.1–0.2	0.2–0.4	0.1–0.2
6. Restoration of eroded soil	0.5–1.0	0.2–0.5	0.2–0.5	0.1–0.2
7. Restoration of salt-affected soils	N/A	0.1–0.2	N/A	0.05–0.1
8. Agricultural intensification	0.05–0.01	0.05–0.1	0.2–0.4	0.1–0.2
9. Water conservation and management	0.05–0.1	0.1–0.3	0.01–0.1	0.1–0.3
10. Afforestation	0.2–0.5	0.1–0.3	0.2–0.5	0.05–0.1
11. Improved pasture management	0.20.05	0.1–0.3	0.1–0.2	0.05–0.1
12. Secondary carbonates	N/A	0.0–0.2	N/A	0.05–0.1

returned to the soil. It is estimated that this will take 25–50 years with current technology (Lal et al. 1998). With good management practices, it may be possible to exceed the original native SOM content of many soils.

Lal et al. (1999) suggest that intensification of agriculture on good soils can be achieved through the widespread adoption of:

1. conservation tillage and residue management;
2. irrigation and water management systems; and
3. improved cropping systems, including agroforestry.

5.2.1 Conservation Tillage and Residue Management

Conservation tillage (CT) is a method designed to keep most crop residue on the soil surface. In this way, soil is protected against erosion, and water losses by run-off and evaporation are also reduced. Reliable data show that traditional intensive tillage decreases the amount of soil carbon, as it encourages rapid mineralization of soil organic matter. Fallow periods in rotation have been used in semi-arid regions to conserve moisture for succeeding crops. However, fallowing, especially in combination with conventional tillage, exposes the soil to erosion (especially wind) and creates temperature and moisture conditions that speed up the decomposition of organic matter in the soil.

Conservation tillage is generally considered to store or build more organic matter in the soil and provide long-term productivity and sustainability by enhancement of soil quality and improvement of soil resilience (Reicosky et al. 1995; Grant 1997). Technologies related to conservation tillage have been widely adopted by farmers in North America and are currently gaining momentum with farmers elsewhere. About 37% of the land farmed in the USA is now managed with a CT system, including no-till, minimum-till, and ridge-till methods (Lal et al. 1999). The reduction in costs associated with the application of these methods is an encouraging advantage for the farmers of the area.

Crop residue is one of the most important conservation tillage factors for improving the physical and chemical properties of the soil. Residue helps to reduce surface run-off and soil loss, conserving soil moisture and improving soil micro-organism populations, soil organic matter content, and soil hydraulic/physical properties. The effectiveness of residue is linked to the soil topography and soil slope, as well as to other factors that affect the sustainability of the residue on the soil surface (Iowa State University Extension 2009).

5.2.2 Irrigation and Water Management Systems

Irrigation, especially in drought-prone soils, can enhance SOC content. Experimental data on the impact of irrigation on SOC dynamics is rare. Bruce et al. (1999) suggest that conversion of dryland to irrigated agriculture may increase the SOC content with an average rate of 100 kg/ha/year. Irrigation of soils in arid and semi-arid regions also affects the SOC pool and its dynamics. This is a complex issue and very few attempts are being made to increase our understanding of this aspect of the carbon cycle.

5.2.3 Improved Cropping Systems, Including Agroforestry: Commercial Fertilization

It is obvious that the application of fertilizers (N, P, and K) will increase overall biomass production, including root biomass. Long-term experimental studies around the world have clearly proven this. Fertilizers have been in use, especially for the last 40–50 years, with the aim of increasing food production. We should recognize that this is a good strategy for increasing food production in developing countries, which in turn can itself help to stabilize deforestation and to reduce greenhouse gas emissions. More biomass production means an increased chance of carbon sequestration. This can also be achieved by other organic inputs, crop rotation, and agroforestry.

5.2.4 Organic Fertilization and Other Organic Inputs

This includes greens, particularly of the legume species, manure compost, manure sewage sludge, wood chips, and peat, beside crop residues. Adding organic matter on to severely eroded soils reduces the risk of erosion by promoting the formation of aggregates that resist erosion.

5.2.5 Crop Rotation

Crops have different needs, and they also have different effects on the soil. Farmers rotate the crops so that the plants do not use up all of the nutrients from the soil. Crop rotation works by farmers alternately planting crops which will aid in restoring soil fertility rather than planting only crops which deplete the soil carbon. There are many potential benefits of diversified or complex crop rotations. Crop rotation also controls diseases and insects. Plants, especially vegetables such as tomatoes, cauliflower, and cabbage, planted in the same location each year will encourage the build-up of certain diseases in the soil. By rotating crops, one can prevent the spread of diseases and viruses. Forage grasses and legumes have extensive rooting systems that leave large amounts of organic matter. When used with conservation tillage, crop rotation adds more organic matter to the soil.

5.2.6 Agroforestry

This is a system of combining fast-growing trees with agriculture that includes growing feed to support livestock (Mergen 1986). It provides a habitat for biodiversity and produces goods and services (Winterbottom and Hazelwood 1987).

This system can substantially increase carbon sequestration (Unruh et al. 1993). It is a compromise solution between continuous crop production, supporting livestock, and carbon sequestration.

The World Agroforestry Centre (ICRAF) has been established at Nairobi in Kenya specifically to deal with agroforestry. Little is known about carbon storage within a given time frame (Schroeder 1993). Among the key results one can get from agroforestry are that it can be applied in all regions (though at different levels in different regions, e.g. it is more significant in Central America and less so in East Asia), that tree cover is strongly positively related to humidity, and that there are mixed relationships between tree cover and population density, depending on the region (Zommer et al. 2009).

One of the advantages of agroforestry systems is that they provide a more hospitable environment for biodiversity, both above and below ground. Sanchez et al. (1996) suggest that substantial biodiversity benefits are likely if agroforestry covers a large area and is maintained for a relatively long time.

Estimating the potential for increasing carbon sequestration is still difficult. The flux of carbon among plants, soils, and the atmosphere is not well appreciated. Understanding how to increase soil carbon stocks in agricultural lands is critical to increasing the sustainability of food production and the mitigation of degraded lands, especially in arid and tropical regions.

5.3 Soil Carbon in the Mediterranean Regions

The lands around the Mediterranean are characterized by a wide diversity of soils, reflecting differences in climate, landscape, vegetation, time, and especially the long-term influence of human activities. The effect of anthropogenic intervention on soil organic carbon in these ecosystems is unknown (Munoz et al. 2007).

Erosion has been a dominant factor in shaping landscapes and influencing the distribution of soils. In the history of soil science, one of the best-known examples of Mediterranean soil is the famous "*terra rossa*". However, there are difficulties with the classification of soils that depends on the observation of variable solum thickness, evidence of clay transportation within the profile, carbonate content, age of soil development, and the superimposition of soils representing ancient pedogenesis. Therefore, their classification may vary from Regosols or Leptosols to Luvisols and Vertisols. Understanding the formation and behaviour of Mediterranean soil is a challenge that still requires the thinking of many pedologists (European Commission—Joint Research Centre. Institute for Environment and Sustainability 2012).

Strategies for sustainable soil management in Mediterranean regions include conserving soil organic matter, minimizing erosion, enhancing soil fertility, and balancing production with environmental sustainability, especially in areas of drought. As mentioned above, no-tillage agricultural ecosystems have the potential to encourage carbon flux from the atmosphere to enter the soil, leading to better soil use and management. Carbon sequestration is expected to mitigate global climate change. However, there are many barriers to implementing no-tillage management practices, especially in the South Mediterranean region. The most significant barrier is that developing countries in the South Mediterranean are driven by poverty. Therefore, if no-till soil management is to be used to help address the problem of global warming, priority should be given to implementing policies that alleviate poverty and support small-scale farmers (Mrabet 2010).

The impact of a reduction in precipitation on the soil–atmosphere exchange of net greenhouse gases (including CO_2, CH_4, and N_2) is not well known. Studies in Mediterranean oak woodlands, which cover a wide area in the Iberian Peninsula, of intra- and inter-annual carbon balance and variation in CO_2 fluxes by Pereira et al. (2007) showed that seasonal drought and drying-re-wetting cycles are a major feature in determining the soil–CO_2 exchange. Studies by Shvaleva et al. (2011) in evergreen oak woodlands in southern Portugal confirmed the concept that seasonally dry ecosystems (Mediterranean) may represent a significant sink for atmospheric CH_4. The study further provided evidence that a decrease of 26% or an increase of 10% in the ambient rainfall with an annual precipitation of about 500 mm did not significantly affect soil functionality and had a limited impact on the soil–atmosphere net GHGs.

The Mediterranean region can be considered as a transitional zone between subtropical and temperate climates, characterized by dry, hot summers and mild, wet winters. According to the FAO database, the prevailing soil type in the Mediterranean region is Cambisols, but Fluvisols, Luvisols, and Leptosols are also quite common. Soils in the Mediterranean region demonstrate a wide variety of parent materials, drainage conditions, and topography. A large body of information is now available on soil carbon and carbon sequestration in the Mediterranean region (Jandl et al. 2011). The Mediterranean region is one of the most densely inhabited areas of the world and the ecosystems have been exploited from the time of ancient cultures to the present.

Through the COST 639 action programme, a project was developed to quantify the soil carbon content of the major Mediterranean ecosystems within the collaborating countries. The contribution of wildfires to soil carbon loss and organic matter decomposition were investigated, together with CO_2 soil emissions. The average soil carbon content to a depth of 30 cm was found to be around 60–70 t C/ha^{-1} for forest and rangeland ecosystems. Lower contents were found for agricultural soils. However, surprisingly high values, up to 200 t C/ha^{-1}, were also reported, which seems to be quite uncommon for dryland ecosystems (Rodeghiero et al. 2011).

Andreetta et al. (2011), by studying the humus for forms of soil carbon storage in the Mediterranean forest soils in southern and central Italy, found that a clear-cut relationship existed between humus form and SOC stock in particular intervals and in the sites they studied. They suggested that this was due to the dominant influence of the nature and genesis of the A horizon. They concluded that Mull humus represents nutrient-rich systems, with a strategy of fast nutrient cycling. Assuming that the dynamics of nutrients and organic carbon are linked, and turnover for organic C is fast, the large carbon stocks of their "Mull-macro" forms would have resulted from the high net primary ecosystem productivity. They also concluded that humus forms have a clear potential as indicators of organic carbon status in Mediterranean forest soils.

Persiani et al. (2008) have studied the relationship between soil fungi species and soil biodiversity and soil carbon storage. Their study has provided the first empirical evidence of natural patterns of soil fungal biodiversity and soil carbon storage in the Mediterranean grasslands at different elevations. They observed an asymptotic relationship between morphological and functional trait richness and fungal biodiversity. Even with a limited number of fungal species, it was possible to attain morphological and functional variability in the soil fungal assemblages in the grasslands studied. They found that increasing fungal biodiversity and biochemical specialization were related to higher soil carbon storage (at a higher altitude and lower slope position), even in fenced-off grasslands.

5.4 Examples of Carbon Dynamics Under Different Land Use Systems

5.4.1 Example from: Temperate Soils of Canada (Acton and Gregorich 1995)

Canada has two main national objectives in the matter of carbon cycling. These are

1. to determine whether Canadian agriculture is a source or sink of atmospheric CO_2, and
2. to reduce uncertainties about the processes that determine the exchange of CO_2 between land and atmosphere.

Quantitative methods were designed to measure carbon stored in the soil under different management technologies and CO_2 release from the cultivated soils. It was calculated *that the amount of CO_2 released from the soil, when native Prairie grasslands were first cultivated, was equivalent to that released by about 10 years of fossil fuel consumption in Canada.* Currently there seems to be a balance established with the farming system used. Several methods were tested to sequester atmospheric CO_2 in soils and several management options were also identified in Canada.

The quantification of carbon storage requires many years of studies, as the altered new system attains a balance between soil carbon inputs and outputs only after so many years. This is unfortunately one of the problems faced by carbon sequestration studies. Canada is interested in developing land management systems that maintain biodiversity, sustainability, and agricultural competitiveness. There is a desire to eliminate summer fallow and use the land to grow legumes as a green manure. Canadian ecological conditions allow only one crop a year, and the

Table 5.3 Organic matter at two depths after 18 years of various tillage treatments of a soil from Ontario, Canada under corn. *Source* Acton and Gregorich (1995)

Tillage system	Soil organic matter (t/ha)		
	0–15 cm	15–30 cm	0–30 cm
No-till	86	65	151
Chisel plough	73	52	125
Disc	74	58	133
Moldboard plough	66	64	130

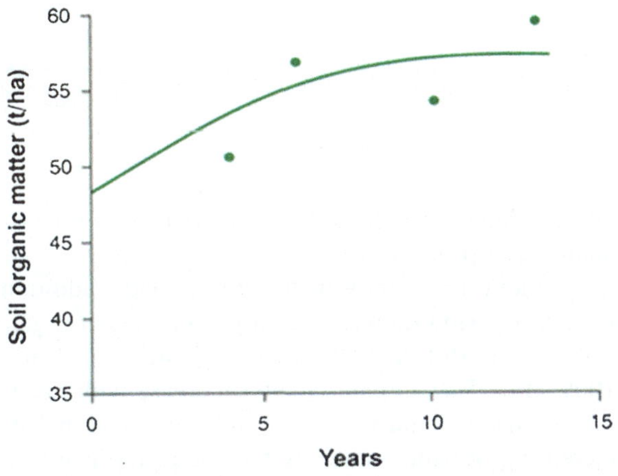

Fig. 5.2 Effects of no-till on soil organic carbon within the top 15 cm of the soil. *Source* Acton and Gregorich (1995)

growing season is short due to low atmospheric and soil temperatures. The climate varies from semi-arid to humid.

About 18 years of tillage treatments of soils from eastern Canada under corn showed that no-till increased organic matter in the soil (both at the surface and throughout the soil profile). Table 5.3 shows clearly that organic matter is increased with no-till treatment.

A study in Saskatchewan also demonstrated the steady increase of soil organic matter when soil was used for conservation tillage (Fig. 5.2). After 10 years or so, it seems that organic matter content has reached a steady state. It was found that the rate of increase varied with regions and soils. The system that uses conservation tillage plus fertilization helps to conserve organic matter in all soils in Canada.

Using computer simulation models, changes in soil organic matter within the top 15 cm of the soil layer of virgin soils subjected to 50 years of conventional tillage, conservation tillage (or no-till), and no-till plus fertilization with nitrogen at an annual rate of 50 kg/ha were predicted. As can be seen in Fig. 5.3, conventional till without fertilizers predicted further decline in organic matter content. The rate of decrease slows down with the treatment no-till without fertilizers. However, with conservation tillage and adequate fertilization, the model predicted a significant increase in soil organic matter over present-day levels, suggesting that

Fig. 5.3 Soil organic matter levels as predicted by computer simulation models under three different management systems. *Source* Acton and Gregorich (1995)

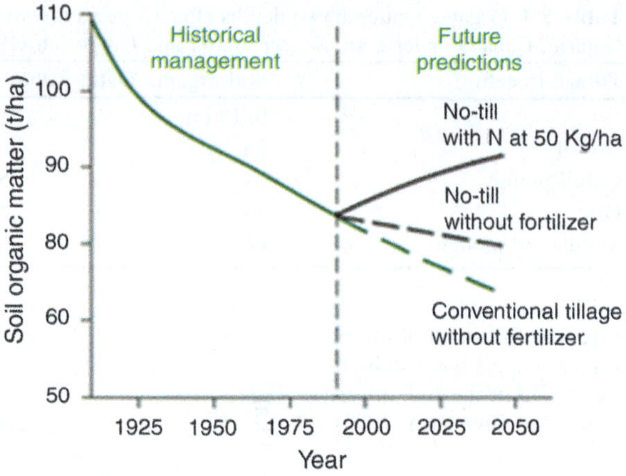

this combination will sequester carbon with time and consequently increase soil quality and productivity.

In a long-term study in Alberta, Canada, addition of manure as applied over a fifty-year period showed that large amounts of organic matter can be placed in the soil and the difference between manure-treated and non-treated soil was almost 100% (Fig. 5.4). Organic matter increase with manure exceeded the original. The human-made (anthropogenic) soils in Western Europe are typical examples of increasing organic matter by the use of farmyard manure.

A long-term study of corn grown in Ontario showed the positive effect of manure application on soil organic matter and crop yields, especially when corn

Fig. 5.4 Changes in soil organic matter over 50 years, with and without manure, in Alberta, Canada. *Source* Acton and Gregorich (1995)

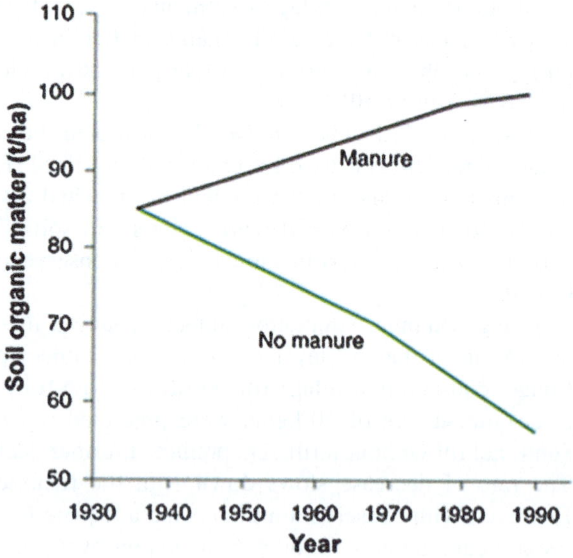

Table 5.4 Organic matter present in the surface horizon (20 cm) cropped with corn and under different management schemes. *Source* Acton and Gregorich (1995)

Crop	Soil organic matter tonnes per ha	Corn grain yield tonnes per ha
Continuous corn		
Fertilized	97	6.0
Unfertilized	88	
Corn in rotation		1.6
Fertilized	112	7.8
Unfertilized	88	4.6

was rotated with other crops (Table 5.4). Use of forage crops works well in eastern Canada where there is a livestock industry, but this is not practical where livestock production is limited.

References

Acton, D. F.; Gregorich, L. J., 1995: *The Health of Our Soils: Toward Sustainable Agriculture in Canada* (Ottawa: Agriculture and Agri-Food Canada, Research Branch. Center for Land and Biological Resources Publ. 1906/E).

Andreetta, A.; Ciampalini, R.; Moretti, P.; Vingiani, S.; Poggio, G.; Matteucci, G.; Tescari, F.; Carnicelli, S., 2011: "Forest Humus Forms as Potential Indicators of Soil Carbon Storage in Mediterranean Environments", in: *Biol Fertil Soils*, 47: 31–40.

Bruce, J. P.; Frome, M.; Haites, E.; Janzen, H.; Lal, R.; Paustian, K., 1999: "Carbon Sequestration in Soils", in: *Journal of Soil and Water Conservation*, 54: 382–389.

European Commission—Joint Research Centre. Institute for Environment and Sustainability, 2012: *European Soil: A Global Perspective*, European Commission Joint Research Centre. Institute for Environment and Sustainability: Land Management and Natural Hazard Unit.

Grant, F. R., 1997: "Changes in Soil Organic Matter under Different Tillage and Rotations: Mathematical Modeling in Ecosystems", in: *Soil Science Society of America Journal*, 61: 1159–1175.

Iowa State University Extension, 2009: *Residue Management and Cultural Practices* (Ames: Iowa State University Department of Agricultural and Biosystems Engineering).

Jandl, R.; Rodeghiero, M.; Olsson, M., (Eds.), 2011: *Soil Carbon in Sensitive European Ecosystems: From Science to Land Management* (Hoboken, NJ: Wiley-Blackwell).

Kapos, V.; Ravilious, C.; Campbell, A.; Dickson, B.; Gibbs, H. K.; Hansen, M. C.; Lysenko, I.; Miles, L.; Price, J.; Scharlemann, J. P. W.; Trumper, K. C., 2008: *Carbon and Biodiversity: A Demonstration Atlas* (Cambridge, UK: UNEP-WCMC).

Lal, R.; Follet, R. F.; Kimble, J. M.; Cole, V. R., 1999: "Managing US Cropland to Sequester Carbon in Soil", in: *Journal of Soil and Water Conservation*, 54: 374–381.

Lal, R.; Kimble, J. M.; Follet, R. F.; Cole, V. R., 1998: *The Potential of U S Cropland to Sequester Carbon and Mitigate the Greenhouse Effect* (Ann Arbor, MI: Sleeping Bear Press).

Mergen, F., 1986: "Agroforestry—an Overview and Recommendations for Possible Improvements", in: *Tropical Agriculture*, 63: 6–9.

Mrabet, R., 2010: "Climate change and Carbon Sequestration in the Mediterranean Basin: Contributions of No-Tillage Systems", in: *Options Méditerranéennes*, A no., 96: 165–184.

Munoz, C.; Ovalle, C.; Zagal, E., 2007: "Distribution of Soil Organic Carbon Stocks in an Alfisol Profile in Mediterranean Chilean Ecosystems", in: *Journal of Soil Science Nutrition*, 7,1: 15–27.

Olson, D. M.; Dinerstein, E.; Wikramanayake, E. D.; Burgess, N. D.; Powell, G. V. N.; Underwood, E. C.; D'Amico, J. A.; Itoua, I.; Strand, H. E.; Morrison, J. C.; Loucks, C. J.; Allnutt, T. F.; Ricketts, T. H.; Kura, Y.; Lamoreux, J. F.; Wettengel, W. W.; Hedao, P.; Kassem, K. R., 2001: "Terrestrial Ecoregions of the World: A New Map of Life on Earth", in: *BioScience*, 51,11: 933–938.

Pereira, J. S.; Mateus, J. A.; Aires, L. M.; Pita, G.; Pio, C.; David, J. S.; Andrade, V.; Banza, J.; David, T. S.; Paço, T. A.; Rodrigues, A., 2007: "Net Ecosystem Carbon Exchange in Three Contrasting Mediterranean Ecosystems—the Effect of Drought", in: *Biogeosciences,* 4: 791–802.

Persiani A. M.; Maggi, O.; Montalvo, J.; Casado, M. A.; Pineda, F. D., 2008: "Mediterranean Grassland Soil Fungi: Patterns of Biodiversity, Functional Redundancy and Soil Carbon Storage", in: *Plant Biosystems*, 142:,111–119.

Reicosky, D. C.; Kemper, W. D.; Langdale, G. W.; Douglas, C. L., Jr; Rasmussen, P. E., 1995: "Soil Organic Matter Changes Resulting from Tillage and Biomass Production", in: *Journal of Soil and Water Conservation*, 50: 253–262.

Rodeghiero, M.; Rubio, A.; Díaz-Pinés, E.; Romanyà, J.; Marañón-Jiménez, S.; Levy, G., et al., 2011: Soil Carbon in Mediterranean Ecosystems and Related Management Problems. In Soil Carbon in Sensitive European Ecosystems: From Science to Land Management (Wiley 2011: 176–218.

Sanchez, P. A.; Buresh, R. J.; Leakey, R. R. B., 1996: Trees, Soils and Food Security. Paper Presented at the "Discussion Meeting on Land Resources: On the Edge of Malthusian Precipice?", London, 5 December 1996.

Schroeder, P., 1993: "Agroforestry Systems: Integrated Land Use to Store and Conserve Carbon", in: *Climate Research*, 3: 53–60.

Shvaleva, A.; Lobo-do-Vale, R.; Cruz, C.; Castaldi, S.; Rosa, A. P.; Chaves, M. M.; Pereira, J. S., 2011: "Soil–Atmosphere Greenhouse Gases (CO_2, CH_4 and N_2O) Exchange in Evergreen Oak Woodland in Southern Portugal", in: *Plant Soil Environment*, 57,10: 471–477.

US Department of Energy, 1999: *Carbon Sequestration, State of the Science* (Washington: A Working Paper for Roadmapping Future Carbon Sequestration Research and Development. US Department of Energy Office of Science, Office of Fossil Fuel).

UNEP-World Conservation Monitoring Center, 2009: *Annual Report 2009* (Cambridge, UK: UNEP-WCMC).

UNEP, 1997: World atlas of desertification, Second Edition, N. Middleton and D. Thomas (eds.) UNEP New York.

Unruh, J. D.; Houghton R. A.; Lefebvre, P. A., 1993: "Carbon Storage in Agroforestry: An Estimate for Sub-Saharan Africa", in: *Climate Research*, 3: 39–52.

Winterbottom, R.; Hazelwood, P. T., 1987: "Agroforestry and Sustainable Development: Making the Connection", in: *Ambio*, 16: 100–110.

Zommer, R. J.; Trabucco, A.; Coe, R.; Place, F., 2009: *Trees on Farm: Analysis of Global Extent and Geographical Patterns of Agroforestry* (Nairobi, Kenya: World Agroforestry Centre) ICRAF Working Paper no. 89.

Chapter 6
Soil and Terroir

Edoardo Antonio Costantino Costantini and Pierluigi Bucelli

Abstract This review aims to draw together the main traits of an interdisciplinary research field that bridges different branches of pedology with crop sciences and food analysis. It introduces the concept of *terroir*, placing a special emphasis on research that has investigated the role played by soil on terroir recognition. As the availability of water and oxygen are the main drivers of the effect of soil on terroir, a case study is reported that demonstrates the interactions between the soil water regime and vine phenology and evapotranspiration. Other soil functional qualities are reported for wine grapes as well for olive oil, apples, oranges, pomegranates, potatoes, carrots, truffles, beer, cheese, coffee, and tobacco. Soils of the well-acknowledged terroirs are often characterized by a moderate supply of water, oxygen, and nutrient, are probably regulated by a specific biodiversity, and are able to ensure that the target qualitative result is achieved without massive integration of fertilizers. Thus a challenge for soil scientists is to provide farmers with the evidence and the tools to enable them to select, maintain, and carefully support soils of terroir. A special focus is then provided on the new frontiers that soil sciences can offer in this field of research, and in particular on the new materials and methods that can be used to understand and measure the soil influence on wine grape and other quality crops. Advances are demonstrated in: (1) the use of proximal and remote soil sensors, models, and statistical analysis; (2) the spatial and temporal assessment of nutrient availability and soil biology; (3) the adoption of the carbon isotope ratio to assess the stress suffered by the plant during the growing season; and (4) new methodologies to trace the origin of food.

Keywords Precision agriculture · Mediterranean · Vis-NIR · NDVI · Kriging · Metagenomic · $\delta^{13}C$ · Strontium

---

E. A. C. Costantini (✉) · P. Bucelli
Consiglio per la ricerca e la sperimentazione in agricoltura,
CRA—Research Centre for Agrobiology and Pedology,
Piazza DAzeglio 30, Florence 50121, Italy
e-mail: edoardo.costantini@entecra.it

S. Kapur and S. Erşahin (eds.), *Soil Security for Ecosystem Management*,
SpringerBriefs in Environment, Security, Development and Peace 8,
DOI: 10.1007/978-3-319-00699-4_6,

6.1 Introduction

Quality crops produce foods that are appreciated for their quality and in particular their taste. It is well known that the taste of food can be affected by a number of causes, such as technological process, cultivar, and agro-technique, but there is a growing awareness that the physical environment of production plays a great role in food taste. The study of the relationships between the qualities of the physical environment and the taste of a product is evidence for the existence of a local positivity of the resources closely connected with the specificity of the environmental characteristics. Such specificity is generally indicated with the term "cultural vocation" and its valorization is considered one of the major aims to be pursued for the success of modern agriculture in the global market.

The great variability of physical environments can be seen as a constraint on the production of foods with uniform taste characteristics and which could be easily recognizable worldwide, but it may also be considered as an opportunity to increase the "peculiar taste" of each territory. In fact, the singularities of land components and their functionality in crop quality, when properly known and managed, make a tract of land "exclusive". This produces in consumers the awareness that what they are consuming is the fruit of a "unique" combination of natural land elements and human skills. This awareness adds a non-material value to the food and fosters public acknowledgement of the "vocation" of a territory.

On top of that, the valorization of the crop vocation of a territory can be one of the more effective tools for the protection of quality and typicity of foods, as well as the protection of soil from the risks of its degradation. Awareness that the quality of a food is closely linked to a particular environment creates the basis of a general concern for the preservation of the land characteristics and qualities of that environment. In addition, acknowledging that the farmers and the population of a territory are particularly careful in the preservation of the land qualities necessary to the achievement of a high quality of food adds further value to the image of the agricultural product on the market.

Although already known and formalized in ancient Egypt and Roman classical times, the modern recognition of the link between soil, climate, man, variety, and the quality of a food originated in France, and was applied to wine a couple of centuries ago (Vaudour 2003). This led to the formulation of the concept of "terroir", which has only recently been formalized by the International Organisation of Vine and Wine (OIV 2010). Nowadays, besides wine grapes, there are a number of other crops for which the "terroir effect" has been demonstrated.

Soil is unanimously considered a major component of terroir, and so this review deals first with experiences in investigating the role played by soil characteristics and qualities in the terroir effect. It then focuses on the new frontiers of pedology in this field of study, and in particular on the new materials and methods that can be used to understand and measure the soil influence on wine grape and other quality crops.

6.2 The Concept of Terroir

The current official definition of terroir says: "Vitivinicultural 'terroir' is a concept which refers to an area in which collective knowledge of the interactions between the identifiable physical and biological environment and applied vitivinicultural practices develops, providing distinctive characteristics for the products originating from this area. 'Terroir' includes specific soil, topography, climate, landscape characteristics and biodiversity features" (OIV 2010) (see Fig. 6.1).

When applied to all crops, a terroir can be defined as a tract of land whose natural characteristics, namely soil, subsoil, relief, and climate, form a unique assemblage of factors which give to the agricultural product, through plants or animals, specific and high-quality characteristics. Man has geared agricultural husbandry and processing technologies to the particular natural environmental conditions to enhance the quality of food and confer on it distinctiveness and exclusivity

A terroir is a delimited area whose size varies according to natural and cultural factors, and which allows an economic dimension to the production (Vaudour 2002, 2003; Deloire et al. 2005; Van Leeuwen and Seguin 2006). Although the extent can vary considerably, many terroirs span some tens of hectares (Bodin and Morlat 2003; Morlat and Bodin 2006; Costantini et al. 2006). Terroirs can be mapped at detailed or semi-detailed scales (mainly from 1:5,000 to 1:25,000), delineating lands with homogenous climatic, topographical, geological, pedological, and managerial characteristics, with acknowledgment of their origin and typicity (Bonfante et al. 2011). The terroir concept can not only be related to an area of already consolidated production, but can also be applied to a territory of recent development. As a rule, the terroir concept is introduced and popularized through the creation of an initial terroir with the acknowledgment of high-quality status within a potentially suitable territory. This implies the need for somewhat high initial investment.

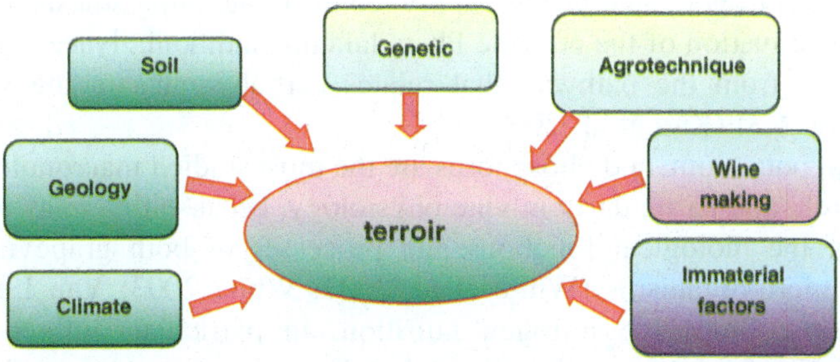

Fig. 6.1 Factors of terroir. *Source* The authors

6.3 Some Mechanisms Governing the Soil Effect on Terroirs for Wine Grape

Soil is considered a major component of terroir, and most evidence that relates wine to the specific soil conditions of the vineyard is empirical (Van Leeuwen et al. 2004). It has been largely demonstrated, for instance, that the best terroirs for red wine production are often situated where some soil limiting factors reduce vine vigour and berry size (Van Leeuwen and Seguin 2006), so that grapes ripen completely but slowly. These limiting factors may be chemical (poor nitrogen nutrition or ion antagonisms, for example) or physical (insufficient water supply during certain phases of the vegetative cycle of the wine).

As a whole, nitrogen nutrition and water supply during certain phases of the vegetative cycle of the vine are considered essential factors of wine quality. Their role in determining the terroir effect has been experienced in many wine-producing areas and with several varieties, such as in France with Cabernet Sauvignon (Choné et al. 2001), Merlot (Trégoat et al. 2002), and Sauvignon Blanc (Peyrot des Gachons et al. 2005); in Australia with Sauvignon Blanc (White et al. 2007); in Hungary with Kékfrankos (Zsófi et al. 2009); and in the USA with Cabernet Sauvignon and Chardonnay (Chapman et al. 2005; Deluc et al. 2009).

In addition to empirical evidence, a few mechanisms have been understood. In particular, it has been proved that soil water availability influences the hormonal equilibrium of each vine variety, which in turn regulates the expression of the genotype (Champagnol 1997; Van Leeuwen and Seguin 1997).

During the early stage of growing, the rate of shoot elongation follows the concentration of auxins, gibberellins, and cytokinins (Fig. 6.2). After flowering, there is a parallel decrease in the quantity of growth hormones and shoot elongation. Veraison is characterized by the increase of ethylene and abscisic acid, and the consequent rate of berry growth. This continues until a threshold at which berry growth sharply decreases and juice concentration starts. The further secretion of ethylene and abscisic acid induces plant ageing, which ends at the stage of the technological maturity of the grape, and harvest.

Similarly, nitrogen and water supply control the biosynthesis of flavonols through the activation of the enzyme Phenylalanine ammonia lyase, which diverts phenylalanine from the pathway that relates carbohydrates to the synthesis of proteins (Fig. 6.3) (Kao et al. 2002).

Nitrogen, potassium, and phosphorus are the most studied macronutrient in soil-plant research, and particularly in vine physiology, because they play a major role in many of the biological functions and processes of both grapevine and fermentative micro-organisms (White et al. 2007; White 2003; Van Leeuwen and Seguin 2006). Grapevine nitrogen nutrition, in particular, influences quality components in the grape and, ultimately, the wine. In addition, fermentation kinetics and formation of flavour-active metabolites are also affected by the nitrogen status of the must. Plant uptake of soil nitrogen increases berry concentration of the major nitrogenous compounds, such as total nitrogen, total amino

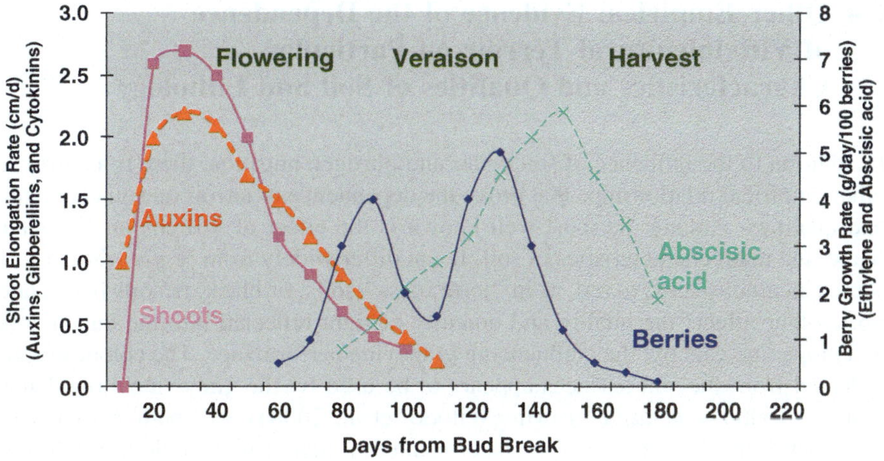

Fig. 6.2 Hormonal equilibrium and vine phenology. *Source* Adapted by the author from Fregoni (2005)

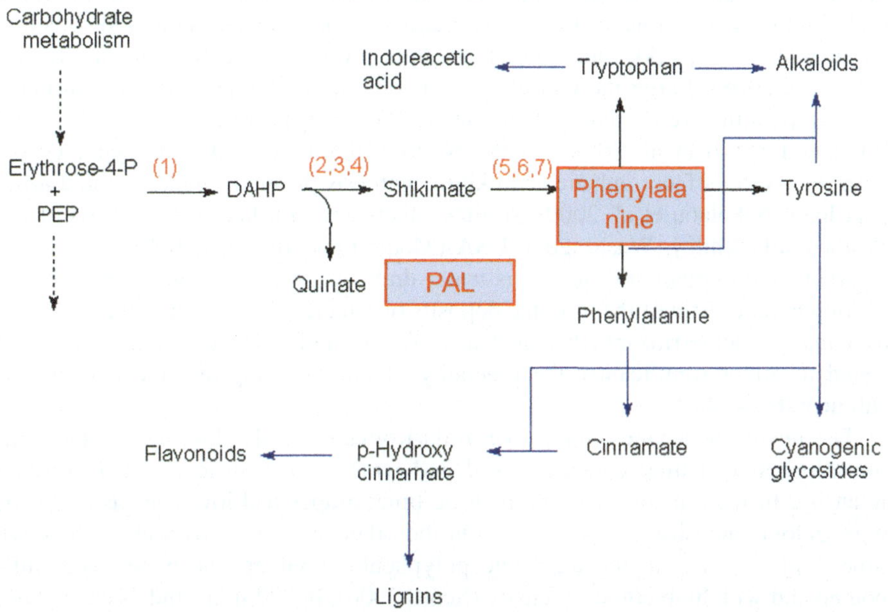

Fig. 6.3 Mechanism of biosynthesis of flavonols. *Source* Adapted by the author from Kao et al. (2002)

acids, arginine, praline, and ammonium, and consequently *yeast-assimilable nitrogen* (YAN). Intermediate must YAN favours the best balance between desirable and undesirable chemical and sensory wine attributes (Bell and Henschke 2005).

6.4 Other Empirical Evidence of the Dependence of Vitivinicultural Terroir on Particular Characteristics and Qualities of Soil and Lithology

In addition to the influence of soil water and nitrogen nutrition, there is an array of other empirical relationships that prove the dependence of terroir on soil characters and qualities. Among the most well-known is the effect of soil colour. Colour is one of the main characteristics of soil. It can differ widely from bright white, as for some calcareous soils, to red, as in "terra rossa" soils, or black, in soils from slate. Soil colour affects the quality and quantity of light reflected into the bunch zone and grapevine canopy, thus influencing grapevine performance. The colour of light reflected from the soil surfaces appears to be used by the grapevine's regulatory system to alter vegetative growth (Witbooi et al. 2008a). In South Africa, with Cabernet Sauvignon, grey soil surface resulted in higher grape colour at 520 nm, while the potassium content of the pulp was the highest with red soil surface treatments (Witbooi et al. 2008b). Stony soils reflect heat if they are pale-coloured. Well-known examples are the white cobbles of Chateauneuf-du-Pape and the pebbles at Sancerre (France) and at Monsant (Spain). In contrast, the metamorphic rocks and grey limestone of Franconia (Germany) provide dark-coloured soils that warm relatively quickly and store heat, thus promoting ripening in this region (Maltman 2008). Large thermal effects on soil surface temperature and on berry skin temperature were found at Geisenheim (Germany) for the vines Riesling and Pinot noir (Stoll et al. 2008). In the North Willamette Valley (Oregon, USA), basalt-derived surfaces enhance cytokinin synthesis through spreading the diurnal heat load (Nikolaou et al. 2000); similar effects arise further north in parts of the Walla Walla Valley, Washington, USA (Meinert and Busacca 2000).

Another important soil quality is water drainage, which is considered a major terroir characteristic of the moraine deposits of Fanciacorta (northern Italy) as well as of many other territories (Panont et al. 1997). A rapid soil water drainage has been found to affect significantly the precocity of bud-breaking and the intensity of summer stress.

The results of a vine zoning in Emilia-Romagna (Italy) highlighted the relationship between lime content in soil and wine colour, structure, and perfume intensity. In fact, in soils with no or little lime, grapes had lower levels of sugar, wine colour intensity, and structure. On the other hand, in soils with high active lime, grapes had a higher sugar and polyphenols content, and wines were full-bodied and with high colour intensity (Scotti 2006). In addition, studies carried out in Alto-Adige (northern Italy) with Schiava vines demonstrated that total polyphenols of grape increased with the increase of active lime in soil (Fregoni 2005).

The potassium content of soil can have a strong effect on must acidity. In particular, the vine responds to an over-absorption of potassium by synthesizing malic acid to neutralize the surplus of K^+ ions. The reaction determines the decrease in acidity and the increase in pH (Hale 1977; Calò et al. 2002). Berry potassium content has been also related to the pH value of must and some authors have suggested that any factor that reduces the photosynthetic activity of leaves could increase potassium accumulation in berries (Freeman et al. 1982). Such factors can be water stress, wind exposure, and excessive shading of the canopy, as with Cabernet Sauvignon in South Africa (Carey et al. 2008).

The most recent research pays attention to micronutrients rather than macronutrients. In particular, sulphur is an essential constituent of the amino acids cysteine and methionine involved in the protection of tissues against oxidative stress (Kopriva 2006), and is involved in the vine's defence strategy against certain fungal and bacterial pathogens (Cooper and Williams 2004). Calcium is important in maintaining membrane integrity. Its deficiency causes cell walls to disintegrate and membranes to become leaky, which causes cell death (Hirschi 2004). Zinc has a useful role in protecting the plant against many environmental stresses (Apel et al. 2004). Manganese was found correlated with phenolics in grape berries (Bramley and Janik 2005); they found that maps of the concentration of phenolics in the berries at vintage and of manganese in petioles at flowering were very similar. Natural abundance of those micronutrients is probably somewhat related to the determination of the soil effect on terroir.

High soil salinity strongly affects vine performance. Excess of salt has both an osmotic and a toxic effect, causing reduction in yield, shoot growth, and berry weight (Lanyon et al. 2004). However, Costantini et al. (2009) demonstrated better performance of Sangiovese in the Chianti area when cultivated on very fertile but moderately saline soils, when the salinity was confined to the deep soil horizons.

The geology of a region is also deemed to be an important component of terroir (Vaudour 2003; Maltman 2008). Geology influences the shape of the landscape, conferring morphology, typical spaces, and articulations that characterize a production district. Geology also influences the morphology of a territory, and thus the climate of vineyards, through altitude, the aspect of the slope, vicinity to water bodies, and exposure to dominant winds. At Stellenbosch (South Africa) in particular, it has been demonstrated that site differences in wind exposure have a stronger effect on Sauvignon blanc than seasonal climatic differences (Carey et al. 2008).

The nature of rock governs deep drainage, but also the quality of groundwater and irrigation water. A relevant characteristic of rock is the degree of its resistance to root penetration. This property derives from rock type, the presence of planes of weakness, and their spacing and orientation (Myburgh et al. 1996). For instance, some of the best grapes produced in the Upper Douro area of Portugal, as well as in Priorat in Catalunya (Spain), Languedoc Roussillon (southern France), and Chianti (central Italy), are obtained from shallow soils on clay schist ("galestro" in Italian). The foliation of schist provides surfaces for root penetration in an otherwise impenetrable material.

Furthermore, there is a common assumption that the kind of rock or sediment determines the physical and chemical composition of the soil of the vineyard (White 2003).

6.5 The Role of Terroir in Timing Vine Phenological Phases

The character of the terroir dictates the phenology of the cultivar. Every variety is identified by its genetic characteristics, and these influence its vegetative, productive, and qualitative behaviour, factors that are more or less stable in different environments. Many parameters are determined mostly by genetic patrimony, and particularly by the trends of phenological phases or the basic characteristics of the cluster and the must (phenolic and aromatic quality and richness, sugar/acid ratios, etc.). However, different varieties have different reactivity and stability regarding the cultivation site. In fact, for grapevines, there are varieties that adapt well even in dissimilar environments, always guaranteeing satisfying results (for example, Cabernet Sauvignon and Chardonnay, grown worldwide), which interact in a minor way with the surrounding environment, compared to vine species that instead readily respond even to small pedological and climatic variations, such as Sangiovese (Paliotti et al. 2008). The work of soil scientists may be decisive regarding this second genetic type, avoiding the risk of declassing varieties and/or environs following bad evaluations. In fact, the vine species can fulfil their potential only if placed in a suitable area, or on the other hand an environment may stand out and gain renown only if a suitable vine species is cultivated. The same considerations may be made for clones and their behaviour.

In the most suitable soils for a certain vine variety, high-quality wines are obtained every year, in spite of climatic variations (Seguin 1986). Actually, it is especially in poor years that the superiority of the grand cru becomes most apparent, that is, the soil attenuates the harmful effects of extreme climatic conditions such as long drought or heavy rainfall. On top of that, such soils will require minimal chemical addition, an additional factor in optimizing sustainability (Mackenzie 2011).

Much viticultural zoning has indicated that each single environmental situation can be evaluated in terms of potentiality regarding the agronomic reference model and the environmental characteristics that allow its creation. In other words, the oenological result corresponds to a model of growth and maturation of the plant determined by agricultural practices, the climate, and the soil conditions. The evaluation of soils must be carried out in relation to the distance between the specific conditions and the reference conditions, in other words, as a function of the hindrances that the natural conditions oppose to reaching the agronomic objective.

In the Mediterranean environment, the phenology and production potential of crops are determined most of all by water availability, both in terms of water available to the plants and of potential, in other words the force with which soil retains the water. As well as this, vegetation and reproduction of the grapevine, which renews a good part of its absorption roots each year, is deeply influenced by the rate of water available during the year. The phenological reference model for the grapevine is characterized by break of dormancy in the spring, with vegetative resumption lasting until flowering, after which plant growth slows and then stops after the veraison. During ripening, growth ceases and there are slight symptoms of stress. These springtime kinetics of the aerial part of the plant correspond to a strong deepening and absorption of the root system, which is followed by a progressive decrease in activity of the superficial, fine roots that are most responsible for plant water supply and vigor.

The hormonal equilibrium that regulates the timing of phenological phases is determined by the competition between hormones produced in the root extremities (Bahrun et al. 2002). In spring when water is abundant, production of cytokinins is high and plant growth with it. Cytokinins have been demonstrated to be implicated in a long-distance signal that communicates nitrogen availability from the root to the shoot via the xylem (Kudo et al. 2010). Cytokinins are also involved in primordial branching of grapevine clusters (Srinivasan and Mullins 1980) that affects cluster size and thus berry numbers.

At the beginning of summer, corresponding with the flower-setting period, water becomes less abundant and is withheld by soil more tenaciously, in such a way that the ryzosphere might not be able to moisten even at night. Under these conditions, synthesis of cytokinins decreases until it ceases, while the production of abscisic acid is induced, first by the roots, then by the adult leaves (Zhang and Davies 1987, 1989). Thus plant growth is precociously curtailed, in a way that ripening may be compared to a senescence process, characterized by a hormonal equilibrium in the vegetation and in the fruits. Water shortage also induces the accumulation of anthocyanins, tannins, and phenolic compounds in the grape berries, while the degradation of malic acid is favoured: processes that contribute to determining the quality of red wines, in particular their fineness and typicality (Champagnol 1997; Lebon et al. 2003). However, if the conditions of water stress occur too early and too deeply, particularly between bud-breaking and flowering, they hinder foliage development and completion of floral organs. The plant is not able to adapt to the hormonal change and the accumulation of compounds in the fruits occurs in an unbalanced way (Champagnol 1997; Van Leeuwen et al. 1997). Thus, before veraison, water deficit has direct, immediate, and sometimes irreversible effects on the quantity and overall quality of production, since the grape during this phase is involved in an intense activity of cellular multiplication, fundamental for its final size, whereas after veraison, cell expansion takes place, and controlled conditions of slight water deficit not only do not compromise the quality and weight of the grape, but may even lead to a concentration of solutes and a maturation directed more toward the elaboration and accumulation of compounds essential to the quality of the grape, such as aromatic fineness (Poni 2007).

6.5.1 A Case Study

The variety Sangiovese has been the object of a wealth of studies dealing with the relationships between soil and wine (see, for example, Storchi et al. 2005; Costantini et al. 2012), therefore it can be considered a reference to demonstrate the performance of red berry vines in different pedoclimatic conditions.

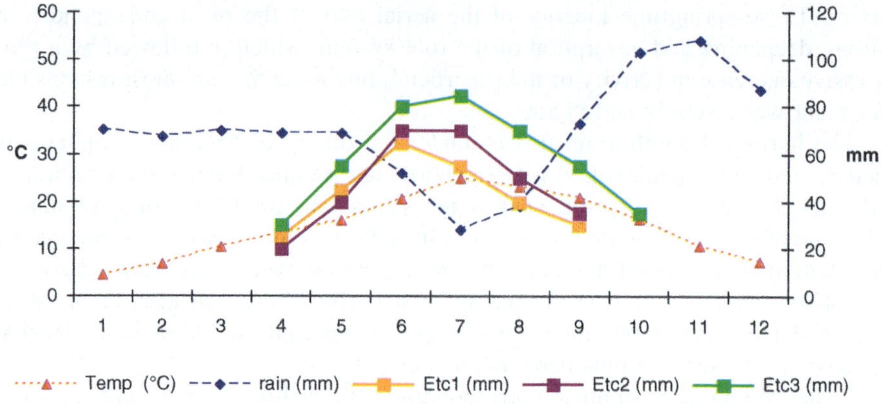

Fig. 6.4 Long-term temperature and rainfall and mean values of water consumption by evapotranspiration of Sangiovese in three different soils of the province of Siena (central Italy). *Source* The authors

A long-term experimental study conducted in the province of Siena (Tuscany, central Italy) demonstrated that the viticultural and oenological performance of Sangiovese depends on both year-dependent and independent pedoclimatic variables (Costantini et al. 2008). Figure 6.4 summarizes the behaviour of Sangiovese cultivated in three soils placed in the same climatic and geological conditions.

The evapotranspiration of vine[1] differs according to the hydrological characteristics of the three soils, as well as the gravimetric amount of unavailable[2] and available water,[3] and air capacity[4] (Costantini et al. 1996).

Cusona soils belong to the Haplic Regosols (Calcaric Arenic) Reference Soil Group (FAO, IUSS, ISRIC 2006) and have a coarse texture, rapid drainage, and low *available water capacity* (AWC). Water consumption through evapotranspiration is prompt and reaches the maximum monthly value in June, but vines suffer from drought in most years during veraison and ripening (Fig. 6.4, Etc1). The hydrological monitoring of the soil reveals abundant oxygen availability throughout the year, but lack of available water in the first 75 cm during August and early September (Fig. 6.5). Because of the prolonged water stress, accumulation of sugar and polyphenols in berries slows down to a different extent according to the length and intensity of the water stress, while the malic acid is consumed instead. The oenological result is therefore much dependent on the climate, and it is often characterized by wines that are not well-balanced or suitable for ageing (Fig. 6.6).

---

[1] Estimated with the Epic model (Costantini et al. 2002).

[2] Held at tension greater than conventional wilting point, 1,500 kPa.

[3] Held at tension between wilting point and field capacity, 1,500 and 33 kPa.

[4] Total porosity from bulk density minus pores filled by water at field capacity.

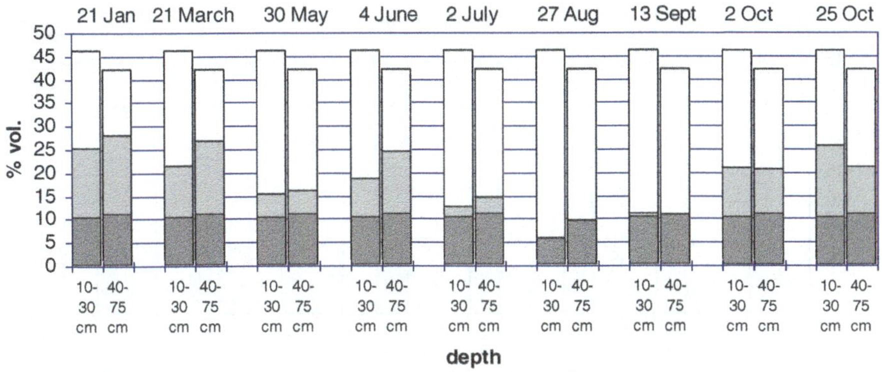

Fig. 6.5 Hydrological behaviour of a Cusona soil: air (*upper part of the bar*), available water (*middle*), and unavailable water (*lower bar*). *Source* The authors

Fig. 6.6 Cusona soils on marine sands are thin and poorly structured, prone to water erosion; vines suffer from prolonged summer water stress. *Source* The authors

San Gimignano soils (Haplic Cambisols (Calcaric)) are loamy and well-structured. They have a fairly good amount of available water and air throughout the entire crop season (Fig. 6.7). Evapotranspiration is greatest and reaches its maximum in August, but then decreases slowly, since the plant continues vegetating until October (Fig. 6.4, Etc3). The vines tend to produce much biomass, even if pruned. Grape bunches are too big and berries too large, so that sugar and polyphenols content is lower than optimal. Instead, total acidity is too great. The resulting wine is little structured, not suitable to ageing, and almost every year penalized at the tasting panel (Fig. 6.8).

The clayey Quercia soils (Vertic Cambisols (Calcaric Siltic)) have well-structured topsoil but the subsoil tends to be poorly aggregated. Root penetration is limited in depth because of the lack of oxygen, especially in springtime. During

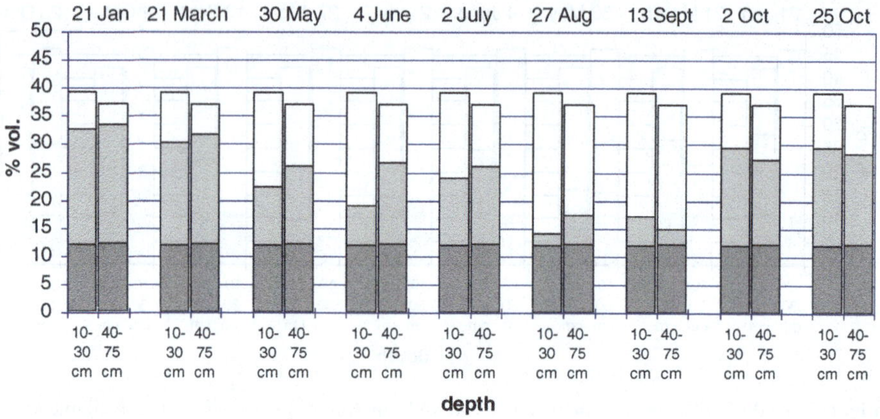

Fig. 6.7 Hydrological behaviour of a San Gimignano soil: air (*upper part of the bar*), available water (*middle*), and unavailable water (*lower bar*). *Source* The authors

Fig. 6.8 San Gimignano soils on marine and colluvial sands are very thick, well-structured, and fertile; vines show excessive vegetation and yield. *Source* The authors

summer, roots must elongate in the lower horizon to extract water, which is rather strongly held in the micropores (Fig. 6.9). This condition of moderate stress and limited and slow water uptake corresponds to the phenological model of Sangiovese, with a well-balanced vegetation/bunches ratio, small berry size, rich in sugar and polyphenols, ensuring good oenological potentiality (Bucelli et al. 2010) (Fig. 6.10).

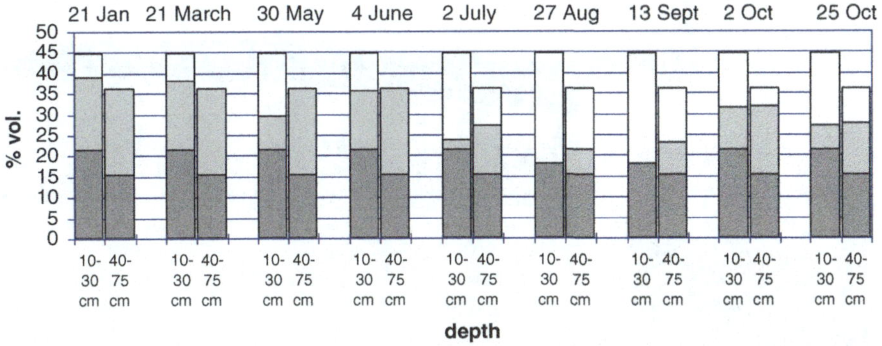

Fig. 6.9 Hydrological behaviour of a Quercia soil: air (*upper part of the bar*), available water (*middle*), and unavailable water (*lower bar*). *Source* The authors

Fig. 6.10 Quercia soils on marine clays are moderately deep, with slight hydromorphism and salinity in depth; vines show moderate stress after veraison. *Source* The authors

Monte soils (Stagnic Regosols (Calcaric, Hyposodic)) are similar to Quercia ones, but lack the subsurface cambic horizon as a consequence of excessive bulldozing and earth movement. It is the activity carried out to prepare the fields for the plantation of tree crops, namely deep ploughing and slope reshaping, which may cause the outcropping of the salt-affected layers, and the failure of cultivation (Fig. 6.11). Hydrological monitoring of this soil reveals very harsh conditions for vine growth (Fig. 6.12). The oxygen deficiency in most of the profile during early spring, and related colder temperatures, cause delayed vegetative development of the roots, as well as of the shoots. In this way, when the summer water climatic deficit starts, roots have not elongated enough to adsorb water from the soil mass, and vines suffer the most.

In spite of the overall negative effect of an excess of salts on many crops, it has been demonstrated that a moderate salinity of the deeper soil horizons of the profile can enhance the oenological outcome of some vine cultivars (Costantini

Fig. 6.11 Monte soils on marine clays are poorly structured and drained, and slightly salty at shallow depth; vines show the strongest stress. They form as a consequence of excessive earth movement before the plantation of a vineyard. Salt efflorescence's and dead vines in the foreground, leopard-like spots of Sodic soils in the background vineyard. *Source* The authors

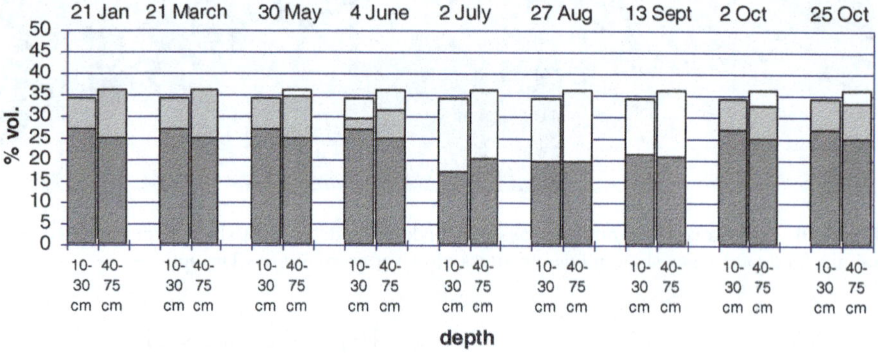

Fig. 6.12 Hydrological behaviour of a Monte soil: air (*upper part of the bar*), available water (*middle*), and unavailable water (*lower bar*). *Source* The authors

et al. 2010; Scacco et al. 2010). In particular, a moderate salinity in the subsoil causes a limited physiological water stress during the phase of berry ripening, which limits the grape yield and improves the quality of the must. This can be crucial for vine varieties that tend to an excess of vegetation when they are cultivated on soils that have a large available water capacity, and hence a good water supply even during the driest part of the growing season.

6.5.2 Terroirs of Other Quality Crops

The "terroir concept" has become popular in many parts of world. Originally developed for wine, it is now applied to many other quality crops such as fruits, vegetables, olive oil, coffee, cacao, and other food products. In fact, besides wine grape, there are a number of other crops for which the terroir effect has been demonstrated.

Olive oil is surely a food where quality plays an important role. Studies have linked soil and land parameters to the qualitative result of olive production. In Italy, Lulli and collaborators demonstrated the terroir effect in Calabria and in Sardinia (Lulli 2004), Tittarelli and others in Latium (Tittarelli et al. 2002), and Bucelli and collaborators in the province of Siena (Bucelli et al. 2011). Soil functional factors for olive cultivation were addressed for both site and profile characteristics. In the former case, elevation and aspect were taken into consideration: elevation was acknowledged very suitable when it was between 200 and 400 m above sea level, and not suitable where it was more than 800 m above sea level. As for the aspect, south, south-west, and east–west were the most suitable. Experimental results showed that the olive tree root system needs a good balance between water and air capacity, and so soils with good or moderate drainage were more suitable. Soils with poor drainage were prejudicial unless the difficulty in draining was reduced by good external drainage (run-off).

Soil hydrology was also demonstrated to be functional for olive trees by Chaves et al. (1976) and Franchini et al. (2009). The *available water capacity* (AWC), in particular, was the parameter that most regulated olive tree results, both quantitatively and qualitatively.

Research conducted in Tuscany (central Italy) was carried out to verify the influence of soil water availability on nutraceutical components and the sensory profile of the virgin olive oil obtained from rain-fed olive trees (*Olea europea*) of the varieties "Frantoio" and "Moraiolo" (Bucelli et al. 2011). Soil moisture was monitored during climatically contrasting years in two olive groves, both having plants of the two cultivars which were homogeneous in age, density, and agricultural husbandry. The soils were a Skeleti Calcaric Regosol and a Haplic Calcisol (FAO, IUSS, ISRIC 2006) with similar chemical characteristics, but contrasting hydrological properties. The plots had the same morphological and climatic conditions. Sixteen monocultivar oil samples were analyzed for fatty acids, *minor polar compounds* (MPC), and tocopherols, and submitted to the organoleptic analysis of a panel of trained tasters. The results emphasized the role played by the soil water regime in the nutraceutical components and sensory evaluation of the olive oil produced. The soil which had a relatively more intense and longer water deficit during summer (Regosol) had an earlier ripening and gave the best results in terms of MPC, in particular oleuropein aglycon, hydroxytyrosol, tyrosol, and luteolin, and, consequently, anti-oxidant properties of the olive oil. The sensorial properties of the oil obtained from both cultivars on the Regosol were superior in both years of trial. Thus, the soil water regime, along with the climate of the year, were the most important factors affecting the nutraceutical components and sensorial quality of virgin olive oil.

In China, Chai et al. (2009) studied the relationship between water and soil environment and apple quality. They addressed the major factors affecting Fuji apple quality from groundwater and soil quality, in a case study carried out in Qian and Baishui counties. The soil and water mineral element concentration and apple quality traits were tested and analyzed. The result indicates that groundwater and soil composition were closely correlated with apple quality. Mg, pH, Cu of groundwater and available K and pH of soil showed strong correlation with total acid, sugar and acid ratio, and total sugar. From the results, the Fuji apple suitability map, based on soil and water conditions, was achieved.

In China again, Zhang (1966) made a comparison between the nutrients of different soil parent materials from which oranges with various qualities were obtained, and demonstrated that oranges with the finest quality occurred in a particular area that had complex mineral compositions. There was a complex relationship between orange quality, Mn, Mo, Zn, and Ca content, and the ratios Ca/K and Mo/N.

The relation between pomegranate quality and the mineral nutrition in soils was studied in Yuanshi County, Hebei Province (China) (Gao et al. 2008). The results showed that the quality of sweet pomegranate planted in biotite-plagioclase gneiss was better than that planted in soils formed on other rocks, and in soils with well-developed cracks. The higher P, Mn, Zn, and Sr content and lower Ni and Cr content were correlated with the higher quality of pomegranate planted in biotite-plagioclase gneiss. Meanwhile, though the K content of biotite-plagioclase gneiss was lower, it mainly occurred in the delay-effective state and was easily absorbed by pomegranate. According to an analysis of soil elements, the distribution of high-quality pomegranate was related to the distribution of P, Mn, Zn, and Sr.

The relationship between soil and potato quality (variety Agria) was studied in Calabria (southern Italy) in the district of Sila Grande over five years of research (Lulli et al. 2009). Three soil series were compared: Cecita soil series on fluvial-lacustrine deposits, made up of gravel, silt, and clay; Sila series evolved on granite, granite-diorite, mica-schist, and gneiss; and Croce della Palma series on granite, granite-diorite, quartz-diorite, mica-schist, and gneiss. Potato showed different quality-quantity characteristics as a function of the soil series. In fact, the soils in the Cecita area produced tubers of medium texture and graininess, with low moisture, a fairly strong typical taste, and a chestnut aftertaste, compared to the soil of the Sila series. The Croce della Palma series also exhibited high variability of organoleptic characteristics: the pulp varied from slightly tender to slightly firm, the graininess between slightly fine and fine, and typical taste ranged from slight to strong. Another soil effect was on tuberization, which occured first in the Sila series, then in the Cecita, and last in the Croce della Palma series. Soils in Sila enhanced typical taste and did not exhibit sensations of aftertaste so strong that product quality was compromised. On the other hand, the sample of potatoes coming from the Cecita soils showed a weak typical taste. Lastly, a correlation between the chemical-physical characteristics of soils and potato quality was hypothesized: soils rich in organic matter and with a loam soil texture produced potatoes with an organoleptic profile inferior to those with soils that have a loamy sand texture, with a medium–low content of organic matter.

Terroir and quality of some vegetables was also demonstrated in France (Villeneuve 1997), in particular for carrots, where the size of the root was affected by the texture and structure of the soil. An 'Extra' grade was given to carrots grown in sand. The nutritional value depended on the levels of sugar, vitamin A, ascorbic acid, and fibre, and uneven mineral levels (boron, P, K, Ca, and B). A panel of tasters revealed a clear difference in the organoleptic taste of carrots grown at the Mont-Saint-Michel as compared with those grown in Val-de-Saire, Nantes, and in the Landes region of France. But other factors could interfere with the soil effect, such as the year, the cultivar, the fertilization regime, and the time of harvest.

Among "niche crops", Bragato et al. (2009) reported useful information about the habitats of the five main edible truffle species (*Tuber* spp.) in quite well-defined environments. The edible truffle species require soil softness, good aeration, and the presence of active lime. These conditions can be found in a large range of soil environments, but it is the way soil combines these conditions that plays a decisive role, making a given environment suitable for truffle production. In particular *Tuber magnatum*, also termed white Alba truffle, is present whenever a strong aeration is originated in soil by a noticeable slope or fluvial activity, or pedoturbation due to earthworms, mesofauna, human activity, or other factors. This tuber also requires soil moistening, but does not tolerate water stagnation, so that even when the distance from perennial streams is very short, soils suitable for the fungus are always well drained. *T. melanosporum,* also called Black Norcia or Périgord truffle, needs a somewhat moist soil, summer being the critical season for the truffle growth. The loss of water is often limited by the mulching effect of whitish gravel on the soil surface, or improved by the tuber itself. The suitable sites are located in quite stable landforms, often where soils have undergone a partial decarbonization followed by a re-carbonization. Each truffle species has its own strategy of life, but all of them require moist and draining sub-alkaline soils, characterized by good aeration, a full base saturation, and the presence of carbonates, even if some of their aplotypes tolerate neutral, non-carbonated soils.

Maltman (2004) investigated the effects of geology on brewing. Maltman explained that geology has far more of a direct influence on beer than it does on wine. This is because the chemical properties of the water used for brewing are crucial, and depend on the ground where the water is extracted. Burton-on-Trent (Staffordshire, England), for example, has ideal brewing water, because it has a naturally balanced blend of beneficial ions, perfect for brewing pale ales. Other famous brewing locales such as Pilsen and Budvar (Czech Republic) have suitable groundwater. Dublin is perhaps the most famous case: everyone goes on about how Guinness tastes better when it is made from Liffey water. The water used in Dublin Guinness is taken from the Grand Canal, fed by limestone inputs. This hard water necessitates the use of dark roasted malts, which thus dictate the style of the beer.

For milk and cheese, the word terroir includes physical environments (geology, geomorphology, and climatology), animals, and humans (Grappin et al. 1996; Monnet et al. 2000). For several years, research has been carried out to study the relationship between cheese characteristics and terroir characteristics either using a global approach or through experimental studies on type of forage, breed of cow, etc. An extensive study of twenty Comté cheese plants (France) has shown that there is a close correlation between edaphic and flora characteristics of the production areas and the aroma profiles of the cheeses. The influence of pasture characteristics on the sensory properties of cheese and differences in aroma volatile compounds between mountain and lowlands were demonstrated for different types of cheeses (Abondance and Gruyère). Comté cheese, made from raw whole milk with local starter lactic acid bacteria, has been manufactured in the traditional manner in the French part of the Jura mountains since the thirteenth century. Its flavour varies not only as a function of grass and hay, but also by geographical production sites. Connoisseurs are able to identify the origin (terroir) of a certain number of cheeses. A new method of large-scale cartography (agro-pedological units) covering twenty cheese cooperatives provided a statistical comparison showing the different edaphic sectors. A sensory analysis of 106 cheeses from these twenty cooperatives (summer or winter cheese with three or six months of ripening) has demonstrated Comté terroirs corresponding to 85 % of the edaphic sectors.

These sectors are thus equivalent to cheese terroirs. Terroirs of coffee have been studied in several countries of Central America and North Africa (Montagnon et al. 2006).

In Honduras a survey was conducted in six areas with a total of fifty-two sampled plots (Avelino et al. 2001). This survey allowed the relationship between the environment and the quality of the beverage to be studied through a holistic approach. The data were analysed by a series of multivariate analyses. The results showed that the quality of the beverage was especially connected to the altitude of the plots, rainfall, soil acidity, percentage of shade, yield of the trees, and bean size. The most appreciated, aromatic, and balanced beverages were associated with the plantations located at high altitudes (1,115 m on average), with a medium annual rainfall (1,726 mm), a slightly acid and rich soil, a medium percentage of shade (48 %), a high yield (464 fruit-bearing nodes per coffee tree), and big beans. These results allowed zones of quality coffee production to be identified in Honduras in the areas of El Paraíso, Comayagua, Olancho, and Marcala, but also allowed the quality to be foreseen by taking into consideration the rainfall, the yield, and bean size of the year.

The effect of some soil properties on the productivity and the quality of cherries of coffee in sustainable coffee growth (*Coffea Arabica var. tipica* L.) were studied in Mexico (Comeau and Krasilnikov 2006). This study was developed at the coffee-growing farm of El Sinaí, situated in the eastern part of the Sierra Madre del Sur in the state of Oaxaca, at an altitude of 800–1,300 m. The climate of the region is warm humid isothermals with an annual precipitation of 1,800–2,000 mm and a mean annual temperature of 21–21.9 °C. The coffee was grown under the shade of natural vegetation, classified as Tropical Semi-deciduous Forest. Fifteen sites were selected for this study at each production cycle, and flavour, aroma, body, and acidity of 225 shrubs of coffee (fifteen coffee plants by site with altitude, relief, and well-known localization) were measured. The best soils for coffee plants were deep sandy loam soils with granular structure in the surface horizon, providing good aeration, moderate infiltration, and pH from 5.0 to 6.0. N, P, and K were the most important elements for the coffee plant, but physical conditions were the most important limiting factors.

In Costa Rica two terroirs were examined: Orosi (between 1, 020 and 1, 250 m above sea level) and Santa Maria de Dota (between 1, 550 and 1, 780 m) (Avelino et al. 2005). This study assessed the effects of slope exposure, altitude, and yield on several cup quality criteria of coffees. East-facing slopes gave beverages with generally superior attributes, probably owing to better exposure to morning sunlight. A positive relationship was found between altitude and taster preferences in both terroirs. A negative relationship was also found between yield and beverage acidity at Santa Maria de Dota, where some coffee trees produced up to 13 kg of coffee cherry. Coffees from Orosi were characterized by a floral flavour which depended on slope exposure, whilst coffees from Santa Maria de Dota displayed a chocolate taste, which was more marked at high altitude. In both terroirs the caffeine, trigonelline, fat, sucrose, and chlorogenic acid contents were not well correlated with the sensory characteristics.

A relationship between cup quality of coffee and soil properties was conducted also in the coffee forest ecosystem of south-western Ethiopia (Yadessa et al. 2008). Cup quality of coffee depended on different factors such as the type of coffee, soil conditions, climatic conditions, and processing methods. From seventy-four sample plots, red cherries were hand-picked and dry-processed, and soil samples were also collected. Soil texture, *cation exchange capacity* (CEC), pH, macronutrients and micronutrients were analyzed. The sensorial analysis was carried out by five professional tasters. Results showed that the overall cup quality of wild Arabica coffee was not correlated with total N and available P levels of soil, but significantly and inversely correlated with the N:P ratio. Generally, coffees with better cup quality were those collected from plots with higher levels of soil-available P, K, clay, and silt, but it was inversely correlated with sand content. Higher levels of soil pH, Mg, Mn, and Zn were also associated with improved coffee aroma.

Research conducted by Castelli and Costantini (2009) for nine years in forty-nine plots of flue-cured tobacco around Verona (northern Italy) showed that high soil macroporosity appears to be the main factor for the success of the crop: the best yield results were achieved with a soil air capacity of above 30 %. In terms of physical characteristics, ideal soil will therefore be able to ensure perfect trafficability, optimal oxygenation, good drainage, and deep rooting. Root growth is encouraged by a low amount of available water so that, at the start of field cultivation in spring, the roots are induced to explore a large soil volume, in order to ensure the necessary water supply for the plant during its later development. Soil depth limitations, such as the presence of petrocalcic horizon or peat, were also considered. Therefore, to evaluate the suitability of land for tobacco growing, the focus was placed on functional characteristics, essentially of a physical nature, which were shown to be particularly important for tobacco cultivation and yield.

6.5.3 State of the Art and Future Challenges in Capturing the Soil Influence on Wine Grape and Other Quality Crops

Although it is well recognized that soil is one of the main factors characterizing terroir, it is also well known that soil properties can vary markedly even within a single field, so that a vineyard, for instance, can produce two or more contrasting types of wine. The optimization of agricultural husbandry in relation to soil characteristics is the main focus of Precision Agriculture.

The adoption of methodologies allowing a detailed knowledge of the spatial and temporal variability of soil functional properties is a key factor for Precision Agriculture, along with improvement in the analytical methods that characterize the crop response to the environment.

New breakthroughs are thus needed for the application of the terroir concept to different crops and environments, as well as for new technologies and analytical methods able to capture the soil influence on crop yield and quality.

Promising advances in the adoption of Precision Agriculture are based on: (1) the use of proximal and remote soil sensors, models, and statistical analysis; (2) the spatial and temporal assessment of nutrient availability and soil biology; (3) the adoption of the carbon isotope ratio to assess the stress suffered from the plant during the growing season; and (4) new methodologies to trace the food origin.

The detailed knowledge of soil functional properties needed to adopt Precision Agriculture is difficult to achieve with traditional methods. In fact, traditional monitoring of soil properties is constrained by the number of sites to be sampled by reason of time-consuming and costly field and laboratory work. At present, there is a need to introduce and tune new technologies to reduce survey costs. A strong innovation is constituted by the use of proximal sensors for on-the-fly monitoring of soil characteristics. From this perspective, soil electrical properties (electrical resistivity or conductivity) can be considered as a promising tool, able to represent the basis of complex information for assessing the spatial and temporal variability of many soil physical and chemical properties (e.g. structure, texture, water content, and salinity). Since the method is non-destructive and very sensitive, it potentially offers a very interesting way of describing subsurface properties without digging, with only a few direct observations and laboratory analysis. Taken as a whole, the maps provided by these field-scale sensors are a basis for soil-sampling strategies, as they accurately reflect spatial variation. Additional prospective uses of geophysical survey include assessing the temporal impacts of management on soil and controlling spatial variation of both soil condition and yield potential.

In that regard, the *electromagnetic induction* (EMI) technique (usually using the EM38) is the most widespread geophysical equipment, employed over the past twenty years in agriculture for its easy use and comparatively low cost, although it requires a calibration every time it is used. Moreover, this technique is iron-sensitive, so particular attention needs to be paid in a vineyard. Sometimes the use of invasive geo-electrical instruments is therefore preferred, namely ARP and Veris, which directly inject electrical current into the soil and measure the voltage drop. Their calibration is more stable and less sensitive to errors due to soil heterogeneity, but they are bigger and heavier compared with the EMI sensors.

The measurement of *apparent electrical conductivity* (ECa) is influenced by several factors, including clay content (Doolittle et al. 2002), gravel content (Morari et al. 2009; Priori et al. 2010), soil moisture (Cousin et al. 2009; Tromp-van Meerveld and McDonnell 2009), salinity (Doolittle et al. 2001), clay-pan depth (Sudduth et al. 2005; Saey et al. 2009), etc. As the factors affecting ECa are complex and interrelated, the ECa maps represent a summary of the soil spatial variability in terms of physical (texture, structure, porosity), hydrological (moisture, drainage, clay-pan depth), and sometimes chemical (salinity) features.

Ground-penetrating radar (GPR) is another non-invasive geophysical technique for detecting electrical discontinuities in the subsurface. GPR is a high-frequency electromagnetic method that acquires data quickly and at high spatial density distribution of the desired property (Clement and Ward 2008). The penetration capabilities of GPR are site-specific and depend upon the frequency spectrum of the source excitation signal, the antenna radiation efficiency, and the electrical properties of the subsurface materials (De Benedetto et al. 2012).

The measurement of gamma radiometric emission is another promising tool that has been recently added to the family of soil-proximal sensors. Being sensitive to the low-level radiations naturally emitted by K, Th, and U radioisotopes, the gamma radiometer is able to provide indications of soil water, organic matter, and nutrients within 30–40 cm of the surface (Wong and Harper 1999).

Visible and near infrared (vis–NIR) spectroscopy too is becoming very important in the field of soil-proximal sensing, and many portable vis–NIR spectroradiometers have come on to the market. Vis–NIR spectroscopy can provide much information about the soil in terms of organic matter, texture, moisture content, CEC, calcium carbonate, etc., although it must be properly calibrated (Viscarra Rossel et al. 2009; Stenberg et al. 2010). Libraries containing field or laboratory-collected spectra can be used to develop models to predict target soil properties analysed with routine methods—in particular, soil organic matter (SOM), minerals, texture, nutrients, water, pH, and heavy metals.

Remote sensing and image-processing methodologies to monitor crop variability, and indirectly soil properties, usually make use of airborne or high-detailed satellite multispectral images to elaborate the NDVI (Normalized Difference Vegetation Index) (Pedroso et al. 2010). The NDVI is a numerical indicator that shows the intensity of vegetation vigour, linked to both health status and leaf area. It is the ratio of the *reflectance in the near-infrared* (rNIR) band to the *reflectance of the red in the visible band* (rRED), as follows (Krieger et al. 1969):

$$\text{NDVI} = \frac{(\text{rNIR - rRED})}{(\text{rNIR + rRED})}$$

Chlorophyll, the pigment in plant leaves, strongly absorbs visible light (from 0.4 to 0.7 μm) for photosynthesis. On the other hand, the cell structure of the leaves strongly reflects near-infrared light (from 0.7 to 1.1 μm). The more leaves, the more these light wavelengths are affected.

Consequently, a well-vegetated area in good health will provide low values of reflectance in the red channel and high values of reflectance in the near-infrared channel, which gives a high value of NDVI. On the other hand, in conditions of poor or no vegetation the NDVI index has a low value. The range of variability of the index is between −1.0 and +1.0. Crop NDVI, and therefore crop leaf area is related to the amount of photosynthetically active radiation adsorbed, hence to fruit characteristics and quality (Johnson et al. 2003).

Besides airborne sensors and sensors implemented on satellites, sensors can also be mounted on a micro-UAV (*Unmanned Aerial Vehicle*). Recently UAVs have become much more widely used and economic. Advances in imaging and computation have also made spectral imaging techniques more affordable. By combining these technologies, previously limited to satellite, it is now feasible to monitor crops to obtain frequent and high-resolution data about vegetation and, in perspective, soil conditions. In particular, an attempt has been made to equip a UAV with a multi-spectral imaging system that is able to simultaneously capture three visible and two near-infrared channels (De Biasio 2010).

An alternative approach to the use of proximal and remote sensors for the monitoring of soil water balance and its spatial pattern is based on modelling. There are some examples of empirical or physically-based hydrological models specifically developed or customized to vine plants. The aim is in most cases the assessing of water balance at the vineyard (Lebon et al. 2003; Ramos and Mulligan 2005; Van Leeuwen et al. 2009; Valdes-Gomez et al. 2009; Zhou et al. 2007), or watershed scale (Conan et al. 2003), or the combined simulation of water and nitrogen dynamics (Nendel and Kersebaum 2004). A number of experiments have been explicitly carried out on applying hydrological models to representative soil mapping units for the characterization of terroirs (Costantini et al. 2002; Bonfante et al. 2011). Bonfante and collaborators, in particular, have integrated the estimation of water stress at the landscape scale, obtained through the simulation of crop water status within soil units, with other spatially distributed environmental information, such as soil data, climate, boclimatic index, and potential radiation.

It goes without saying that new statistical methodologies are needed to cope with the wealth of data provided by the different proximal and remote sensors entering the field (Bramley and Janik 2005). The density of the data allows for many descriptors to be calculated over a surveyed area. These descriptors are not

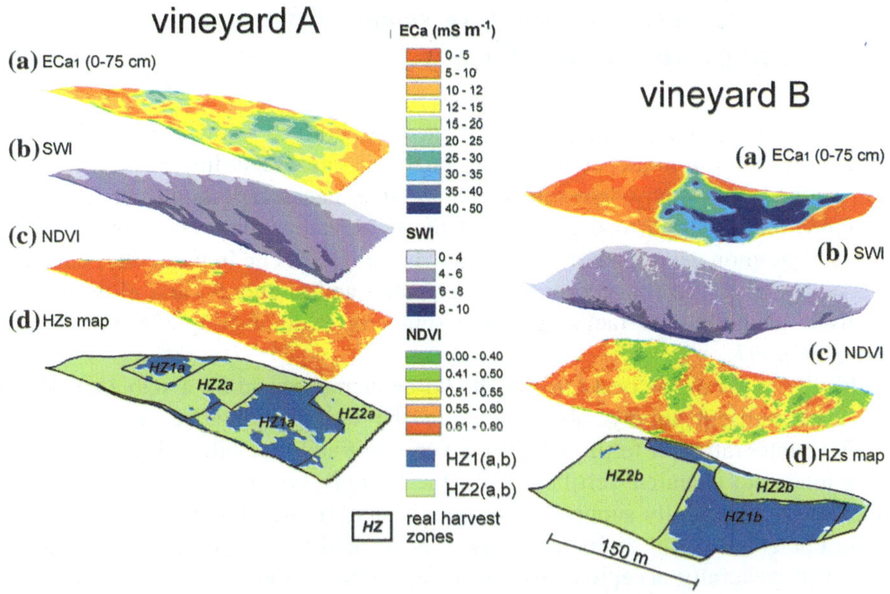

Fig. 6.13 Maps used for the cluster analysis and harvest zones (HZs). ECa is the apparent electrical conductivity; SWI is the *soil wetness index* (from the digital terrain model). NDVI is obtained from satellite image. The *black lines* of the HZ maps show the actual harvest areas, simplified for the manual harvest practices. *Source* The authors

direct measurements of soil properties, but when related to functional soil characteristics they may provide information relevant for discriminating soils, and produce prescription maps and management and harvest zones (Fleming et al. 2000, Costantini et al. 2009).

There are several approaches to combining data from multiple sources in order to make inferences about soil properties. There are various methods for merging primary and secondary information using multivariate statistical and geostatistical techniques. Different multivariate extensions of kriging, such as *kriging with external drift* (KED) (Bourennane et al. 2000; Bourennane and King 2003; Castrignanò et al. 2009), *regression kriging* (RK) (Odeh et al. 1995), *cokriging* (CK) (McBratney et al. 2000), and *factorial kriging analysis* (FKA) and factorial discriminant analysis (Taylor et al. 2009; Morari et al. 2009; De Benedetto et al. 2012). An example of *harvest zones* (HZs) for winegrape obtained by clustering descriptors derived from different sources is reported by Priori et al. (2013) and shown in Fig. 6.13. The wines were produced by the same ordinary processes of the farm cellar and expressed some significant differences between HZs, especially in terms of colour intensity, polyphenols and anthocyanins content, which were used by the farm to differentiate the wine marketing.

6.5.4 Spatial and Temporal Assessment of Nutrient Availability and Soil Biology

Spatial and temporal assessment of nutrient availability is a new frontier of studies in determining the influence of soil on quality crops. According to the Precision Agriculture philosophy, any chemical additive necessary should be precisely designed and metered to every part of the field, rather than added in a blanket 'lowest common denominator' approach (Mackenzie 2011). In particular, reducing the supply of fertilizers according to the spatial and temporal variability of plant needs is an additional factor in optimizing crop sustainability (Davenport and Bramley 2007).

Ideally, soils of the best terroirs should be characterized by a stable and well-balanced nutrient supply, able to assure the target qualitative result without massive integration of fertilizers. Moreover, these soils are often characterized by only moderate chemical fertility. Thus, the farming system should be tailored to maintain and carefully support an agro-ecosystem, rather than to feed a crop.

In this contest, soil biological activity is deemed to play an important role. In fact, it is generally accepted that the mineral composition of the solution transported from the root to the shoot via the xylem, which significantly affects plant development, primarily depends on the interaction between the physical/chemical composition of the soil and the biochemical/biophysical nature of the root systems. On top of that, soil microbial communities surrounding plant roots confer beneficial effects on the plant, such as increased plant growth and reduced susceptibility to disease caused by plant pathogens.

The analysis of soil biological fertility of vineyards can take different forms. The parameters most used are soil organic carbon, carbon/nitrogen ratio, organic carbon mineralization rate, and microbial biomass and respiration (see, for example, Nendel and Reuter 2007). Although a link between soil biology and terroir expression is difficult to assess, some preliminary results indicate that in Mediterranean vineyards better preserved soils are less influenced by environmental variations and show a reduced but more stable biological activity, probably related to a larger biodiversity (Costantini et al. 1996).

In addition to biochemical analysis, the investigation of microbial community structure can be accomplished by sequencing and analyzing the whole DNA directly extracted and purified from soil, the so called "metagenome" (Mocali and Benedetti 2010).

Investigation of microbial community requires bacterial identification and it can be easily accomplished by a metagenomic approach, which may follow both traditional and innovative methods. A preliminary analysis can be carried out by using traditional molecular methods such as *Denaturing Gradient Gel Electrophoresis* (DGGE) in order to detect the more interesting sample sites. In these sites, the whole bacterial community can be assessed by a metagenomic approach, which enables the documentation of microbial diversity by sequencing the PCR amplicons,[5] obtained using universal primers for bacteria 16S rRNA. Amplicons of 16S ribosomal RNA are used to generate a library for *Next Generation Sequencing* (NGS).

As soil microbial communities are key factors for the biogeochemical cycles of both macro- and micronutrients, the determination of functional diversity of microbial communities is also crucial to assessing the best nutrient supply and balance for the best terroir. Therefore the metagenomic approach, which allows us to know the phylogenetic composition of the entire microbial community, should be supported by methods aimed at assessing microbial metabolism and functional diversity. Once again a preliminary analysis could be carried out through the application of the *Community-Level Physiological Profile*, commonly called "BIOLOG" (Garland and Mills 1991). However, the recent development of a new functional gene microarray technology, the Geochip (He et al. 2007, 2011), offers a comprehensive tool for investigating biogeochemical, ecological, and environmental processes. This generation of functional gene arrays contains probes covering approximately 57,000 gene variants from 292 functional gene families involved in carbon, nitrogen, phosphorus, and sulphur cycles, energy metabolism, antibiotic resistance, metal resistance, and organic contaminant degradation. It is particularly useful for providing direct linkages of microbial genes to soil processes and functions related to organic matter mineralization. The proposed methodology will open new insights into a deeper comprehension of soil microbiological functions and soil quality, in order to obtain important indications for optimizing fertilization and soil management.

The study of the soil ecology of terroir should also take into account the mesofauna. In fact, although the relevance of soil biodiversity is widely accepted in ecological terms, interactions with the soil factors regulating food quality are still ignored. Assessing the community of soil micro-arthropod fauna, in particular, might contribute to the knowledge of agro-ecosystem functions such as decomposition, nutrient recycling, and the maintenance of physical-chemical properties such as soil structure (Kladivko 2001). In general, soil invertebrate-based indices consider the consistency and richness of populations. For example, the presence of some soil mite groups can represent information useful for assessing quality in arable farming systems.

Earthworms are another important component of soil mesofauna that might contribute to characterizing different farming systems. Lumbrices and Enchytreids, in particular, affect soil structure, carbon, and nutrient dynamics, both directly and indirectly through their effects on soil micro-organisms. They have been found to differentiate organic viticultural systems in the terroir of Santorini (2006).

Finally it is worthwhile mentioning hydrolytic enzymes, and in particular cellulase and phosphatase, which are produced by organisms in soil to catalyze the

[5] An amplicon is a piece of DNA formed as the product of natural or artificial amplification events. For example, it can be formed via *polymerase chain reactions* (PCR) or *ligase chain reactions* (LCR), as well as by natural gene duplication.

decomposition of large organic molecules such as cellulose and inositol phosphate into smaller monomers such as glucose and phosphate. These enzymes are likely to control the rate of biogeochemical cycling of elements such as carbon, nitrogen, phosphorus, and sulphur in soil, and so they are good indicators of soil biological quality (Dick 1997) and a candidate for terroir indicators.

6.5.5 Carbon Isotope Ratio

Recently, in addition to the more commonly used viticultural indicators, a new physiological marker has been used for describing the vineyard water regime during the ripening period, namely the ratio between the two stable carbon isotopes $^{13}C/^{12}C$, called $\delta^{13}C$, measured in the must sugars upon harvesting, or in the alcohol of the wine produced (Van Leeuwen et al. 2001, 2003; Tregoat et al. 2002; Costantini et al. 2010). The test is based on the carbon isotope discrimination phenomenon that occurs during the course of chlorophyll photosynthesis. The grapevine prefers the absorption of the isotope ^{12}C, because in addition to being much more abundant in atmospheric CO_2 ($\sim$99 %), it is also lighter. Therefore, the sugars formed in the grape have carbon atoms with a higher percentage of ^{12}C isotopes than atmospheric CO_2, and consequently in the grapes the $\delta^{13}C$ ratio is lower. In the case of water deficit stress, isotopic discrimination decreases in intensity. In fact, the leaf stomata close during the hottest part of the day, slowing the photosynthetic activity; the gaseous exchanges between leaf and atmosphere are reduced and therefore also the possibility of discriminating the absorption of the carbon isotopes. In these conditions the $\delta^{13}C$ tends to grow close to that of atmospheric CO_2 (−8 ‰).

The isotopic ratio $^{13}C/^{12}C$ is measured by *Isotope Ratio Mass Spectrometry* (IRMS) and the $\delta^{13}C$ value is expressed as a variable by the thousand (‰) in reference to the international standard V–PDB (*Vienna—Pee Dee Belemnite*). Interest in this proxy is due to the simplicity of sampling: it is sufficient to reserve a representative sampling of the grapes or clusters upon harvesting or during the course of ripening.

For the vineyard the range of values varies between −21.0 and −28.0 ‰ or more, and about −26.0 ‰ was considered the threshold value between water deficit stress and non-limiting water nutrition (Van Leeuwen et al. 2003; Deschepper et al. 2006). In the Chianti area (central Italy), Costantini et al. (2010), using $\delta^{13}C$ in zoning vineyards for the Sangiovese vine, were able to infer that the threshold between good and bad sensory evaluation of wine corresponded to a $\delta^{13}C$ value of -26.7 ± 1.2 ‰. This value might be chosen as a first reference for the delimitation of Sangiovese terroirs.

6.5.6 New Methodologies for Tracing Food Origin

A considerable proportion of typical foods, such as dairy products, olive oil, wine, fruits and vegetables, and cereals are of Mediterranean origin. They are protected by a 'denomination of origin'. The analytical methodologies used for the geographical identification of food products are numerous and for the most part experimental (Karoui and De Baerdemaeker 2007). The most important are DNA analysis and the chromatographic, spectroscopic, and mass spectrometry techniques.

DNA analysis, using the DNA-barcoding methodology, allows a fingerprint for every product to be identified, guaranteeing its origin and quality (Hajibabaei et al. 2007; Kress and Erickson 2008). In fact, DNA sequences, unique to every species or subspecies, can be utilized like a "bar code" to identify a product by comparing them with a database containing the sequences of all known species. Therefore, the identification of so-called molecular markers for the traceability of the agricultural food chain can become a new challenge for the protection of high-quality products.

Among spectroscopic techniques, *nuclear magnetic resonance* (NMR), a physical phenomenon in which magnetic nuclei in a magnetic field absorb and re-emit electromagnetic radiation, provides information on the molecular nature of the product and demonstrates many advantages compared with the classic analytical techniques. It is a non-destructive technique applicable both to liquids (milk, wine, oil, beverages, etc.) and to solids (meat, meals, bread, cheese, etc.) using HR–MAS (*high resolution–magic angle spinning*) drills, without any sample treatments (Sacchi and Paolillo 2006). This technique allows much structural and quantitative information about a sample to be quickly obtained. By matching it with mass spectrometry techniques, it is possible to evaluate the traceability of many typical products.

Besides IRMS, the most commonly-used techniques in the field of mass spectrometry are *high resolution mass spectrometry* (HRMS), *thermic ionization mass spectrometry* (TIMS), and *inductively coupled plasma—mass spectrometry* (ICP–MS).

The study of strontium isotopic characteristics can be an effective method for the geographic traceability of quality wines (Almeida and Vasconcelos 2001). The certification of authenticity can concern the production area and have a legal purpose, but can also be of interest to farms wishing to certify the origin of grapes from a specific region. In fact, the vine soil inherits own characteristics from substrate lithology, so the totality of chemical, organic, and geochemical wine parameters can form its fingerprint and provide a tool of excellent efficacy for traceability. The low concentrations of mineral substances (ppb-ng/g) are dissolved in the wine watery matrix as cations and anions of geological origin, absorbed by plant roots with nutritive elements and fixed inside the berry structure. Some earthy-alkaline elements such as strontium (Sr) and rubidium (Rb) are very important (Boari et al. 2008). The reference concentrations are detectable by means of instrumentation of very high sensitivity. Although the concentrations are very low, their relative distribution is strongly related to the mineralogical composition of soil and substrate rock. The isotopic composition of some of these heavy elements (e.g. $^{87}Sr/^{86}Sr$, $^{143}Nd/^{144}Nd$) certainly derives from the nature and age of the rocks which constitute the vineyard's geological substrate. These characteristics allow the use of the Rb–Sr isotopic to classify both the geological clock for absolute rock dating and the petrogenic marker, since every rock on our planet has a characteristic isotopic $^{87}Sr/^{86}Sr$ ratio (Fortunato 2004; Kelly 2005). Therefore, every soil/rock system where a vineyard is planted has its own isotopic $^{87}Sr/^{86}Sr$ ratio. For instance, a soil on a granite substrate has an isotopic $^{87}Sr/^{86}Sr$ ratio > 0.710 because of the high Rb/Sr ratio of the parent material, which has determined high production of radiogenic ^{87}Sr. On the other hand, soils on basaltic and carbonate sedimentary rock—with a low Rb/Sr ratio—have an isotopic $^{87}Sr/^{86}Sr$ ratio of <0.710. The isotopic $^{87}Sr/^{86}Sr$ ratio is extremely variable from rock to rock, but constant for every rock. Finally, this isotopic ratio could be a real fingerprint of food indicating its origin, and specifying not only the vineyard but also the area of origin (Boari et al. 2008).

Lanthanides or *rare earth elements* (REE) have been studied with particular attention to food traceability (Zhou and Liu 1997; Bontempo et al. 2011). Lanthanides are very interesting for their chemical similarity; they are not subject to selective splitting of concentration from soil to food. So these elements, even if present in small concentrations (ppb or ppt), can be significantly traceable in food, with a distribution corresponding to the underlying soil. This could be especially true for those foods that do not undergo processing before marketing, such as fruit and vegetable products, but it would be interesting to verify this hypothesis for products derived from more elaborate chains such as wine, or from intermediate chains, such as honey. In the field of the wine authentication, there are many studies based on the determination of the elementary outline, but the study of wine traceability is more complex (Augagneur et al. 1996; Thiel et al. 2004). The sample should not have undergone blending, so that the concentrations of the oligo-elements sought depend only on the soil composition of the region of origin. For that, it is necessary to use samples made from a single grape variety and to understand the whole chain from soil to grape, must, and bottled wine. In fact, many studies have shown that the content of lanthanides decreases in absolute terms between the different links of the chain, although their distribution tends to remain constant from soil to must (Rossano et al. 2007). With regard to the change from must to corresponding wine, a strong uncertainty occurs in the evaluation of element distribution, certainly connectable to the lowest concentrations of the same lanthanides in the wines (ng/l of fractions), in particular those at the greatest atomic weight (Bertoldi et al. 2009).

Analysis of lanthanides was also applied to hazelnut, obtaining interesting results (Oddone et al. 2009). Hazelnuts resemble mushrooms and truffles as regards the chain, because there is no intermediate processing between harvesting and consumption of the product, except possible roasting. However, in hazelnuts the concentration levels of lanthanides are lower, apparently because the fruit is in a part of the plant that is far from the soil. On the other hand, the preliminary results show significant differences in lanthanides content in hazelnuts harvested in different areas.

6.6 Conclusions

The best terroirs are rooted in soils that in most years produce a moderate yield but with good quality, without massive integration of fertilizers and unsustainable risks to the preservation of soil quality. Centuries of trial and error have led European farmers to select terroirs as the best pieces of land for a specific crop, and to tune agricultural husbandry towards the target qualitative result. Then farming systems have been tailored to maintain and carefully support the agro-ecosystems that have been created by this harmonic synergy between nature and man.

Nowadays, a shortcut to the selection of land and the tailoring of land management aimed at developing new premium terroirs is only possible by means of a thoroughly and detailed knowledge of the different components of terroir, and this gives a pivotal emphasis to soil.

Spatial variation of soils is often great, especially in Mediterranean countries, through natural as well as anthropic causes. Setting up methodologies and technologies capable of detailed mapping of soil functional properties at reasonable cost and in a reasonable time is certainly one of the main objectives of research in this field. Much effort is currently being expended to better integrate different kinds of proximal and remote soil sensors, both with each other and with other factors controlling the state of the soil (topography, climate, vegetation).

Along with variations in space, knowledge of the modifications of soil characteristics and qualities with time is also essential. The natural seasonality of the climate, particularly pronounced in the Mediterranean countries, interacts with cultivation, fertilization, tractor passage, and irrigation to change soil characteristics and nutrient status during the growing season. The soil monitoring activity should be able to control the spatial variations of functional characteristics with a particular frequency, corresponding at least to the most important crop phenological phases.

Spatial–temporal monitoring of soil water status and oxygen availability at a very detailed scale is still an unresolved issue, especially at high water tension and in stony and frequently cultivated soils. Similarly, detailed nutrient monitoring is also difficult to achieve, either in soil or in plants. Here, the need for information denser than that traditionally attained with "wet" laboratory analysis could be met by the use of new technologies such as infrared spectroscopy.

Finally, recent breakthroughs in the study of soil biology have opened new frontiers in our understanding of the mechanisms regulating the effect of soil on terroir. The metagenome approach, in particular, is a candidate for unravelling the biogeochemical cycles of both macro-and micronutrients, and shedding more light on the functional diversity of terroirs.

References

Almeida, C.M.R., Vasconcelos, S.D., 2001: "ICP-MS Determination of Strontium Isotope Ratio in Wine in Order to be Use as Fingerprint of Its Regional Origin", in: *Journal of Analytical Atomic Spectrometry*, 16: 607–11.

Apel, K., Hirt, H., 2004: "Reactive Oxygen Species: Metabolism, Oxidative Stress, and Signal Transduction", in: *Annual Review of Plant Biology*, 55: 373–99.

Augagneur, S., Médina, B., Szpunar, J., Lobinski, R., 1996: "Determination of Rare Earth Elements in Wine by Inductively Coupled Plasma Mass Spectrometry Using a Microconcentric Nebulizer", in: *Journal of Analytical Atomic Spectrometry*, 11, 9: 713–21.

Avelino, J., Perriot, J. J., Pineda, C., Guyot, B., Cilas, C, 2001: *The Identification of Coffee Terroirs in Honduras: Characterisation of the Environment, Crop Management and the Characteristics of Yield. 19ème Colloque Scientifique International sur le Café*, Trieste, Italy, 14–18 mai.

Avelino, J., Barboza, B., Araya, J. C., Fonseca, C., Davrieux, F., Guyot, B., Cilas, C, 2005: "Effects of Slope Exposure, Altitude and Yield on Coffee Quality in Two Altitude Terroirs of Costa Rica, Orosi and Santa Maria de Dota", in: *Journal of Science Food and Agriculture*, 85: 1869–76.

Bahrun, A., Jensen, C. R., Asch, F., Mogensen, V. O, 2002: "Drought Induced Changes in xylem pH, Ionic Composition, and ABA Concentration Act as Early Signals in Field-Grown Maize (Zea mays L.)", in: *Journal of Experimental Botany*, 53: 251–63.

Bell, S. J., Henschke, P. A, 2005: "Implications of Nitrogen Nutrition for Grapes, Fermentation and Wine", in: *Australian Journal of Grape Wine Research*, 11, 3: 242–295.

Bertoldi, D., Larcher, R., Nicolini, G., Bertamini, M., Concheri, G., 2009: "Distribution of Rare Earth Elements in *Vitis vinifera* L. 'Chardonnay' berries", in: *Vitis*, 48, 1: 49–51.

Boari, E., Tommasini, S., Mulinacci, N., Mercurio, M., Morra, V., Mattei, M., Conticelli, S., 2008: "87Sr/86Sr of Some Central and Southern Italian Wines and Its Use as Fingerprints for Geographic Provenance. Proceedings di OIV 2008-31st World Congress of Vine and Wine, 6 p.

Bodin, F., Morlat, R., 2003: "Characterizing a Vine Terroir by Combining a Pedological Field Model and a Survey of the Vine Growers in the Anjou Region (France)", in: *Journal of International Science*. Vigne Vin, 37, 4: 199–211.

Bonfante, A., Basile, A., Langella, G., Manna, P., Terribile, F., 2011: "A Physically Oriented Approach to Analysis and Mapping of Terroirs", in: *Geoderma*, 167–168: 103–117.

Bontempo, L., Camin, F., Marzocco, L., Nicolini, G., Wehrens, R., Ziller, L., Larcher, R, 2011: "Traceability Along the Production Chain of Italian Tomato Products on the Basis of Stable Isotopes and Mineral Composition", in: *Rapid Communications in Mass Spectrometry*, 25,7: 899–909.

Bourennane, H., King, D., 2003: *Using Multiple External Drifts to Estimate a Soil Variable Geoderma*, 114: 1–18.

Bourennane, H., King, D., Couturier, A., 2000: *Comparison of Kriging with External Drift and Simple Linear Regression for Predicting Soil Horizon Thickness with Different Sample Densities Geoderma*, 97, 255–271.

Bragato, G., Gardin, L., Lulli, L., Raglione, M., 2009: "Edible Truffles (Tuber spp.)", in: Costantini, E.A.C. (Ed.): *Manual of Methods for Soil and Land Evaluation* (Science Publishers: Enfield, NH): 254–266.

Bramley, R.G.V., Janik, L.J., 2005: "Precision Agriculture Demands a New Approach to Soil and Plant Sampling and Analysis—Examples from Australia", in: *Communications in Soil Science and Plant Analysis*, 36, 9–22.

Bucelli, P., Costantini, E.A.C., Storchi, P., 2010: "It is Possible to Predict Sangiovese Wine Quality Through a Limited Number of Variables Measured on the Vines", in: *Journal International Science Vigne Vin*, 44, 4: 207–218.

Bucelli, P., Costantini, E.A.C., Barbetti, R., Franchini, E., 2011: Soil Water Availability in Rainfed Cultivation Affects more than Cultivar some Nutraceutical Components and the Sensory Profile of Virgin Olive Oil", in: *Journal of Agriculture Food and Chemical*, 59, 15: 8304–313.

Calò, A., Tomasi, D., Biscaro, S., Costacurta, A., Giorgessi, F., Lorenzoni, A., Menapace, P., Verzè, G., Di Stefano, R., Tosi, E., Benciolini, G., Bertacchini, A., 2002: "Le Vigne Del Soave. Consorzio Tutela Vini Soave e Recioto di Soave", in: *Soave* (VR), Italy.

Carey, V.A., Archer, E., Barbeau, G., Saayman, D., 2008: "Viticultural Terroirs in Stellenbosch, South Africa. II. The Interaction of Cabernet-Sauvignon and Sauvignon Blanc with Environment", in: *Journal International Science Vigne Vin,* 42, 4: 185–201.

Castelli, F., Costantini, E.A.C., 2009: "Tobacco (*Nicotiana tabacum*)", in: Costantini, E.A.C. (Ed.): Manual of Methods for Soil and Land Evaluation, (Science Publishers, Enfield, NH), 211–221.

Castrignanò, A, Costantini, E.A.C., Barbetti, R., Sollitto, D., 2009: "Accounting for Extensive Topographic and Pedologic Secondary Information to Improve Soil Mapping", in: *Catena* 77, 28–38.

Chai, Y.,Yan, B., Shi, X., 2009: "Effects of Water and Soil Environment Situation for Apple Quality in Weibei Plateau", in: *Journal of Irrigation and Drainage*, 28, 1.

Champagnol, F., 1997: "Caractéristiques édafiques et potentialités qualitatives des terroirs du vignoble languedocien". in: *Atti Colloque international "Les terroirs viticoles"*, Angers, 17–18: 259–263.

Chapman, D.M., Roby, G., Ebeler, S.E., Guinard, J.X., Matthews, M.A., 2005: "Sensory Attributes of Cabernet Sauvignon Wines made from Vines with Different Water Status", in: *Australian Journal of Grape Wine Research,* 11: 339–347.

Chaves, M., Troncoso, A., Romero, R., Prieto, J., Linan, J., 1976, "Estudio de los Caracteres Optimos de los Suolos de Olivar en Diferentes Zonas de Andalucia Occidental, in: *Nutricion del olivo*, Sevilla.

Choné, X., Van Leeuwen, C., Chéry, P., Ribéreau-Gayon, P., 2001: "Terroir Influence on Water Status and Nitrogen Status of Non-Irrigated Cabernet Sauvignon (Vitis vinifera): Vegetative Development, must and Wine Composition", in: *S Af J Enol Vitic,* 22, 1: 8–15.

Clement, W.P., Ward, A., 2008, "GPR Surveys Across a Prototype Surface Barrier to Determine Temporal and Spatial Variations in Soil Moisture Content, Chapter 23", in: Allred, B.J., Ehsani, M.R., Daniels, J.J. (Eds.): *The Handbook of Agricultural Geophysics, American Society of Agricultural Engineers* (US: CRC Press): 305–315.

Comeau, L.P., Krasilnikov, P., 2006: "*Effect of Soil Properties on the Quality and Productivity of Coffee in Mountainous Regions of Sierra Madre del Sur (Southern México), 18th World Congress of Soil Science*". July 9–15 (Philadelphia: Pennsylvania, USA).

Conan, C., de Marsily, G., Bouraoui, F., Bidoglio, G., 2003: "A Long-Term Hydrological Modelling of the Upper Guadiana River Basin (Spain)", in: *Physics and Chemistry of the Earth*, 28,4–5, 193–200.

Cooper, R. M., Williams, J. S., 2004: "Elementar Sulphur as an Induced Antifungal Substance in Plant Defence", in: *Journal of Experimental Botany*, 55, 1974–1953.

Costantini, E.A.C., Campostrini, F., Arcara, P.G., Cherubini, P., Storchi, P., Pierucci, M., 1996: "Soil and Climate Functional Characters for Grape Ripening and Wine Quality of "Vino Nobile di Montepulciano", in: *Acta Hort,* 427 ISHS,:45–55.

Costantini, E.A.C., Pellegrini, S., Vignozzi, N., Ciampalini, R., Magini, S., Barbetti, R., 2002: "Using Different Methods for Calibrating Field Characterisation of Soil Hydrological Qualities for the Vine and Olive Tree Zoning", in: M. Pagliai, R. Jones (Eds.): *Sustainable Land Management-Environmental Protection: A Soil Physical Approach. Advances in Geoecology*, 35: 101–114.

Costantini, E.A.C., Barbetti, R., Bucelli, P., L'Abate, G., Lelli, L., Pellegrini, S., Storchi, P., 2006: Land Peculiarities of the Vine Cultivation Areas in the Province of Siena (Italy), with Indications Concerning the Viticultural and Oenological Results of Sangiovese Vine. in:*Italiana Journal of Geoscience* (Boll. Soc. Geo. It.). 6,147–159.

Costantini, E.A.C., Barbetti, R., Bucelli, P., L'Abate, G. Pellegrini, S., Storchi, P., 2008: "Scale Dependence of Soil and Climate Functional Characteristics for Qualitative Sangiovese Vine Production", Proc 31^ OIV Congress Verona, CD-rom computer file", in: *Organisation International Vigne et Vin* (Paris, France).

Costantini, E. A. C., Pellegrini, S., Bucelli, P., Storchi, P., Vignozzi, N., Barbetti, R., Campagnolo, S., 2009: "Relevance of the Lin's and Host hydropedological Models to Predict Grape Yield and Wine Quality", in: *Hydrology and Earth System Sciences,* 13: 1635–1648.

Costantini, E. A. C., Pellegrini, S., Bucelli, P., Barbetti, R., Campagnolo, S., Storchi, P., Magini, S., Perria, R., 2010: "Mapping Suitability for Sangiovese Wine by Means of δ^{13}C and Geophysical Sensors in Soils with Moderate Salinity", in: *European Journal of Agronomy,* 33, 208–17.

Costantini, E.A.C., Bucelli, P., Priori, S., 2012: *Quaternary Landscape History Determines the Soil Functional Characters of Terroir.* Quaternary International, 265, 63–73. doi:10.1016/j.quaint.2011.08.021.

Cousin, I., Besson, A., Bourennane, H., Pasquier, C., Nicoullaud, B., King, D., Richard, G., 2009: "From Spatial-Continuous Electrical Resistivity Measurements to the Soil Hydraulic Functioning at the Field Scale", in: *C.R. Geoscience*, 341: 859–867.

Davenport, J.R., Bramley, R.G.V., 2007: *Western Nutrient Management Conference*, 7 (Salt Lake City, UT) 25–32.

De Benedetto, D., Castrignanò, A., Sollitto, D., Modugno, F., Buttafuoco, G., Lo Papa, G (Ed.), 2012; *Integrating Geophysical and Geostatistical Techniques to Map the Spatial Variation of Clay Geoderma* 171–172: 53–63.

De Biasio, M., Arnold, T., Leitner, R., McGunnigle, G., Meester, R., 2010: "UAV-Based Environmental Monitoring Using Multi-Spectral Imaging Proceedings of SPIE—The International Society for Optical Engineering, 7668.

Deloire, A., Vaudour, E., Carey, V., Bonnardot, V., Van Leeuwen, C., 2005: "Grapevine Responses to Terroir: a global approach", Journal of International Science Vigne Vin, 4: 149–162.

Deluc, L.G., Quilici, D.R., Decendit, A., Grimplet, J., Wheatley, M.D., Schlauch, K.A., Mérillon, J.M., (…), Cramer, G.R., 2009: "Water Deficit Alters Differentially Metabolic Pathways Affecting Important Flavour and Quality Traits in Grape Berries of Cabernet Sauvignon and Chardonnay. *BMC Genomics*, 10, art, 212.

Deschepper, G., Cassassolles, X., Dabas, M., Pernet D., 2006: Complémentarité des Mesures de Résistivité Electrique des Sols et du ΔC13 du moût dans l'étude et la valorisation des terroirs viticoles. VIth International terroir Congress, Bordeaux, 232–236.

Dick, R.P., 1997: "Soil Enzyme Activity as Integrative Indicators of Soil Health", in: C.E., Doube, B., and Gupta, V. (Eds.) *Biological Indicators of Soil Health*; *Pankhurst* (CAB: Wallingford, New York) 121–156.

Doolittle, J.A., Petersen, M., Wheeler, T. 2001: "Comparison of Two Electromagnetic Induction Tools in Salinity Appraisals", in: *Journal of Soil and Water Conservation*, 56: 257–262.

Doolittle, J.A., Indorante, S.J., Potter, D.K., Hefner, S.G., McCauley, W.M., 2002: "Comparing Three Geophysical Tools for Locating Sand Blows in Alluvial Soils of Southeast Missouri", *Journal of Soil and Water Conservation*, 57: 175–182.

FAO, IUSS, ISRIC 2006: "World Reference Base for Soil Resource in World Soil Resource Report" 103 (FAO, Rome, Italy) 132.

Fleming, K.L., Westfall, D.G., Wiens, D.W., Brodahl, M.C., 2000: "Evaluating Farmer Defined Management Zone Maps for Variable Rate Fertilizer Application", in: *Precision Agriculture*, 2, 2: 201–215.

Fortunato, G., Mumic, k., Wunderly, S., Pillonell, L., Bossett, J.O., Gremaud, G., 2004: "Application of Strontium Isotope Abundance Ratios Measured by MC-ICP-MS for Food Authentication", in: *Journal of Analysis Atomic Spectrometry*, 19: 227–234.

Franchini, E., Cimato, A., Costantini, E.A.C., 2009: "Olive Tree (Olea europea L.). In: Costantini E.A.C. (Ed.) *Manual of Methods for Soil and Land Evaluation* (Science Publishers, Enfield, NH, USA) 402–449.

Freeman, B.M., Kliewer, W.M., Stern, P., 1982: "Influence of Windbreaks and Climatic Region on Diurnal Fluctuation of Leafwater Potential, Stomatal Conductance, and Leaf Temperature of Grapevines", in: *American Journal of Enol Vitic*, 33: 233–236.

Fregoni, M., 2005: "*Viticoltura di qualità. Phytoline*, Affi (VR).

Gao, Y., Shen, J., Liu, W., Shi, Y., 2008: "Relationship Between Pomegranate Quality and Geochemical Element Characteristics in Rock and Soil in Mountain Area of Yuanshi County", in: *Chinese Journal of Eco-Agriculture*.

Garland, J.L., Mills, A.L., 1991: "Classification and Characterization of Heterotrophic Microbial Communities on the Basis of Patterns of Community-Level Sole-Carbon-Source Utilization", in: *Applied and Environmental Microbiology*, 57, 8: 2351–359.

Grappin, R., Coulon, J. B., 1996: "*Local production, milk and cheese: some comments. 3èmes rencontres autour des recherches sur les ruminants*" (Paris, France) 4 et 5 december.

Hajibabaei, M., Singer, G.A.C., Hebert, P.D.H, Hickey, D.A., 2007: "DNA Barcoding: How it Complements Taxonomy, Molecular Phylogenetics and Population Genetics", in: *Science Direct*, 23, 4: 167–172.

Hale, C.R., 1977: "Relation between Potassium and the Malate and Tartrate Contents of Grape Berries", in: *Vitis*, 16: 9–19.

He, Z., Gentry, T.J., Schadt, C.W., Wu, L., Liebich, J., Chong, S.C., Huang, Z., (…), Zhou, J., 2007: " GeoChip: A Comprehensive Microarray for Investigating Biogeochemical, Ecological and Environmental Processes", in: *ISME Journal*, 1, 1: 67–77.

He, Z., Van Nostrand, J.D., Deng, Y., Zhou, J., 2011: "Development and Applications of Functional Gene Microarrays in the Analysis of the Functional Diversity, Composition, and Structure of Microbial Communities", in: *Frontiers of Environmental Science and Engineering in China*, 5, 1: 1–20.

Hirschi, K.D., 2004: "The Calcium Conundrum, Both Versatile Nutrient and Specific Signal", in: *Plant physiology*, 136: 2438–1305.

Johnson, L. F., Roczen, D. E., Youkhana, S. K., Nemani, R. R., Bosch, D. F., 2003: "Mapping Vineyard Leaf Area with Multispectral Satellite Imagery", in: *Computers and Electronics in Agriculture*, 38, 1: 33–44.

Kao, Y.Y., Harding, SA., Tsai, CJ., 2002: "Differential Expression of Two Distinct Phenylalanine Ammonia-Lyase Genes in Condensed Tannin-Accumulating and Lignifying Cells of Quaking Aspen", in: *Plant Physiology*. 130: 756–760.

Karoui, R, De Baerdemaeker, J., 2007: "A Review of Analytical Methods Coupled with Chemometrics Tools for the Determination of the Quality and Identity of Dairy Products", in: *Food Chemical*, 102: 621–40.

Kelly, S., Heaton, K., Hoogerwerff, J., 2005: "Tracing the Geographical Origin of Food: the Application of Multi-Element and Multi-Isotope Analyses", in: *Trends in Science and Technology*, 16: 555–67.

Kladivko E.J., 2001: "Tillage systems and soil ecology", in: *Soil and Tillage Research*, 61,1–2: 61–76.

Kopriva, S., 2006: "Regulation of Sulphur Assimilation in Arabidopsis and Beyond", in: *Annals of Botany* 97: 479–95.

Kress, W.J., Erickson, D.L., 2008: "DNA Barcodes: Genes, Genomics, and Bioinformatics", in: *PNAS* 105, 8: 2761–762.

Krieger, F., Malila, W., Nalepka, R., Richerdson, W., 1969: "Preprocessing Transformations and Their Effects on Multispectral Recognition", in: Proceedings of the 6th International Symposium on Remote Sensing of Environment (University of Michigan, USA): 97–131.

Kudo, T., Kiba, T., Sakakibara, H., 2010: "Metabolism and Long-Distance Translocation of Cytokinins", in: Journal of Integrative Plant Biol 52: 53–60.

Lanyon, D.M., Cass, A., Hansen, D., 2004: "The Effect of Soil Properties on Vine Performance", in: CSIRO Land and Water Technical Report No. 34/04.

Lebon, E., Dumas, V., Pieri, P., Schultz, H.R., 2003: "Modelling the Seasonal Dynamics of the Soil Water Balance of Vineyards", in: *Functional Plant Biology*, 30, 6: 699–710.

Lulli, L., 2004: "Il suolo e la qualità dei prodotti. Bollettino della Società Italiana della Scienza del suolo", 53: 21–24.

Lulli, L., Palchetti, E., Vecchio, G., Caruso, A.D., 2009: Potato (Solanum tuberosum L.). In: Costantini, E.A.C., (Ed.) *Manual of Methods for Soil and Land Evaluation* (Science Publishers, Enfield, NH, USA): 197–210.

Mackenzie, D.E., 2011: "Digital terroir" Geoscientific technologies applied to viticultural site characterisation and varietal matching [Online] http://dc269.4shared.com/doc/OurwdHRb/preview.html.

Maltman, A., 2004: "Wine, Beer and Whisky: the Role of Geology", in: *Geology Today*, 19: 22–29.

Maltman, A., 2008: "The Role of Vineyard Geology in Wine Typicity", in: *Journal of Wine Research,* 19, 1: 1–17.

McBratney, A.B., Odeh, I.O.A., Bishop, T.F.A., Dunbar, M.S., Shatar, T.M., 2000: "An Overview of Pedometric Techniques for Use in Soil Survey", in: Geoderma, 97: 293–327.

Meinert, L.D., Busacca, A.J., 2000: "Geology and Wine 3: Terroirs of the Walla Valley Appellation, South eastern" (Washington State, USA). *Geoscience Canada,* 27: 149–170.

Mocali, S., Benedetti, A., 2010: "Exploring Research Frontiers in Microbiology: the Challenge of Metagenomics in Soil Microbiology", in: *Research Microbiology* 161, 6: 497–505.

Monnet, J.C., Berodier, F., Badot, P.M., 2000: "Characterization and Localization of a Cheese Georegion Using Edaphic Criteria (Jura Mountains, France)", in: *Journal of Dairy Science*: 1692–1704.

Montagnon, C., 2006: "Coffee: Terroirs and Qualities", in: Montagnon (Ed.): 172.

Morari, F., Castrignanò, A., Pagliarin, C., 2009: "Application of Multivariate Geostatistics in Delineating Management Zones within a Gravelly Vineyard Using Geo-Electrical Sensors", in: *Computers and Electronics in Agriculture* 68, 97–107.

Morlat, R., Bodin, F., 2006: "Characterization of Viticultural Terroirs Using a Simple Field Model Based on Soil Depth-II, Validation of the Grape Yield and Berry Quality in the Anjou Vineyard (France)", in: *Plant and Soil*, 281,1–2: 55–69.

Myburgh, P.A., Van Zuyl, J.L., Conradie, W.J., 1996: "Effect of Soil Depth on Growth and Water Consumption of Young Vitis Vinifera L. cv. Pinot Noir", in: *South African Journal of Enol Vitic*, 17: 53–62.

Nendel, C., Kersebaum, K.C., 2004: "A Simple Model Approach to Simulate Nitrogen Dynamics in Vineyard Soils", in: *Ecological Modelling*, 177, 1–2: 1–15.

Nendel, C., Reuter, S., 2007: "Soil Biology and Nitrogen Dynamics of Vineyard Soils as Affected by a Mature Biowaste Compost Application", in: *Compost Science and Utilization*, 15, 2: 70–77.

Nikolaou, N., Magdalini, N., Koukourikou, A., Karagiannidis, N., 2000: "Effects of Various Rootstocks on Xylem Exudates Cytokinin Content, Nutrient Uptake and Growth Patterns of Grapevine Vitis Vinifera L. cv. Thompson Seedless", in: *Agronomie,* 20: 363–373.

Oddone, M., Aceto, M., Baldizzone, M., Musso, D., Osella, D., 2009: "Authentication and Traceability Study of Hazelnuts from Piedmont, Italy", in: *Journal of Agricultural Food Chemical,* 9: 3404–408.

Odeh, O.A., McBratney, A.B., Chittleborough, D., 1995: "Further Results on Prediction of Soil Properties from Terrain Attributes: Heterotopic Cokriging and Regression-Kriging", in: *Geoderma*, 67: 215–226.

OIV (International Organisation of Vine and Wine) 2010. Resolution OIV/Viti 333/2010 Definition of vitivinicultural "Terroir". The General Director of the OIV, General assembly Tbilisi (Georgia) 25th June 2010: 1.

Palliotti, A., Silvestroni, O., Petoumenou, D., Vignaroli, S., Berrios, J.G., 2008: "Evaluation of Low-Energy Demand Adaptive Mechanisms in Sangiovese Grapevine During Drought", in: *Journal International Science Vigne* Vin, 42, 1: 41–47.

Panont, C.A., Bogoni, M., Montoldi, A., Scienza, A., 1997: "Improvement of Sparkling Wines Production by a Zoning Approach in Franciacorta (Lombardy, Italy)", In: Acts Colloque international "Les terroirs viticoles", Angers, 17–18 juillet 1996, 454–460.

Pedroso, M., Taylor, J., Tisseyre, B., Charnomordic, B., Guillaume, S., 2010: "A Segmentation Algorithm for the Delineation of Agricultural Management Zones", in: *Computers and Electronics in Agriculture*, 70, 1: 199–208.

Peyrot des Gachons, C., Van Leeuwen, C., Tominaga, T., Soyer, J.P., Gaudillère, J.P., Dubourdieu, D., 2005: "Influence of Water and Nitrogen Deficit on Fruit Ripening and Aroma Potential of Vitis Vinifera L. cv Sauvignon Blanc in Field Conditions", in: *Journal of Science Food Agriculture*, 85: 73–85.

Poni, S., Bernizzoni, F., Civardi, S., 2007: "Response of Sangiovese Grapevines to Partila Root-Zone Drying: Gas-Exchange, Growth and Grape Composition", in: *Scientia Horticolturae*, 114: 96–103.

Priori, S., Costantini, E.A.C., Capezzuoli, E., Protano, G., Hilgers, A., Sauer, D., Sandrelli, F., 2008: "Pedostratigraphy of Terra Rossa and Quaternary Geological Evolution of a Lacustrine Limestone Plateau in Central Italy", in: *Journal Plant Nutr Soil Science*, 171:509–523.

Priori, S., Martini, E., Costantini, E.A.C., 2010: "Three Proximal Sensors for Mapping Skeletal Soils in Vineyards", in: Proceedings of 19th World Congress of Soil Science, (Brisbane: Australia), CD-ROM.

Priori, S., Martini, E., Andrenelli, M.C., Magini, S., Agnelli, A.E., Bucelli, P., Biagi, M., Pellegrini, S., Costantini, E.A.C., 2013: "Improving Wine Quality through Harvest Zoning and Combined Use of Remote and Soil Proximal Sensing". in: *Soil Science Social American Journal*, 77(3). doi:10.2136/sssaj2012.0376

Ramos, M.C., Mulligan, M., 2005: "Spatial Modelling of the Impact of Climate Variability on the Annual Soil Moisture Regime in a Mechanized Mediterranean Vineyard", in: *Journal of Hydrology*, 306, 1–4: 287–301.

Rossano, E.C, Szilágyi, Z., Malori, A., Pocsfalvi, G., 2007: "Influence of Winemaking Practices on the Concentration of Rare Earth Elements in White Wines Studied by Inductively Coupled Plasma Mass Spectrometry", in: Journal Agriculture Food Chemical, 55, 2: 311–317.

Sacchi, R., Paolillo, L., 2006: "NMR for Food Quality and Traceability". In: Nollet, L. M. L., Toldrà F., (Eds.) *Advances in Food Diagnostics*, (Blackwell Publishing).

Saey, T., Simpson, D., Vermeersch, H., Cockx, L., Van Meirvenne, M., 2009: "Comparing the EM38-DD and Dualem-21S Sensors to Depth-to-Clay Mapping", in: *Soil Science Social American Journal*, 73: 7–12.

Sbaraglia, M., Lucci, E., 1994: *Guida all' interpretazione delle analisi del terreno ed alla fertilizzazione*, (Studio Pedon, Roma) 123.

Scacco, A., Verzera, A., Lanza, C.M., Sparacio, A., Gennam, G., Raimondi, S., Tripodi, G., Dima, G., 2010: "Influence of Soil Salinity on Sensory Characteristics and Volatile Aroma Compounds of Nero d' Avola Wine", in: *American Journal Enol Vitic*, 61: 498–505.

Scotti, C., 2006: *Emilia-Romagna: dalla conoscenza del suolo alla qualità del vino*. Il suolo, 1–3.

Seguin, G., 1986: "*Terroirs* and pedology of vine growing", in: *Experientia*, 42: 861–873.

Srinivasan, C., Mullins, M.G., 1980: "Flowering in Vitis: effects of genotype on cytokinin-induced conversion of tendrils into inflorescences", in: *Vitis*, 19: 293–300.

Stenberg, B., Viscarra Rossel, R.A., Mouazen, A.M., Wetterlind, J., 2010: "Visible and Near Infrared Spectroscopy in Soil Science", in: *Advances in Agronomy*, 107, C: 163–215.

Stoll, M., Stuebinger, M., Lafontaine, M., Schultz, H. R., 2008: "Radiative and Thermal Effects on Fruit Ripening Induced by Differences in Soil Colour", in: *VII International terroir Congress* (Nyon).

Storchi, P., Costantini, E.A.C., Bucelli, P., 2005: "The Influence of Climate and Soil on Viticultural and Oenological Parameters of Sangiovese Grapevine Under Non-Irrigated Conditions", in: *Acta Horticulture*, 689, 333–40.

Sudduth, K.A., Kitchen, N.R., Wiebold, W.J., Batchelor, W.D., Bollero, G.A., Bullock, D.G., Clay, D.E., Palm, H.L., Pierce, F.J., Schuler, R.T., Thelen, K.D., 2005: "Relating Apparent Electrical Conductivity to Soil Properties Across the North-Central USA", in: *Computers and Electronics in Agriculture*, 46: 263–283.

Taylor, J.A., Coulouma, G., Lagacherie, P., Tisseyre, B., 2009: "Mapping Soil Units Within a Vineyard Using Statistics Associated with High-Resolution Apparent Soil Electrical Conductivity Data and Factorial Discriminant Analysis Geoderma 153, 1–2: 278–284.

Thiel, G., Geisler, G., Blechschmidt, I., Danzer, K., 2004: "Determination of Trace Elements in Wines and Classification According to Their Provenance", in: *Analytical and Bioanalytical chemistry*, 378, 6: 1630–1636.

Tittarelli, F., Neri, U., Poletti, P., La Certosa, G., Raus, R., 2002: "Monitoraggio dello stato nutrizionale dell'olivo", in: *L'Informatore Agrario*, 44: 37–51.

Trégoat, O., Gaudillère, J.P., Choné, X., Van Leeuwen, C., 2002: "The Assessment of Vine Water and Nitrogen Uptake by Means of Physiological Indicators, Influence on Vine Development and Berry Potential (Vitis vinifera L. cv. Merlot, 2000, Bordeaux), in: *Journal International ScienceVigne Vin*, 36, 3: 133–142.

Tromp-van Meerveld, H.J., McDonnell, J.J., 2009: "Assessment of Multi-Frequency Electromagnetic Induction for Determining Soil Moisture Patterns at Hillslope Scale", in: *Journal of Hydrology*, 368: 56–67.

Valdes-Gomez, H., Celette, F., De Cortazar-Atauri, I.G., Jara-Rojas, F., Ortega-Farias, S., Gary, C., 2009: "Modelling Soil Water Content and Grape Vine Growth and Development with the Stics Crop-Soil Model Under Two Different Water Management Strategies", in: *Journal International Science Vigne Vin*, 43, 1: 13–28.

Van Leeuwen, C., Seguin, G., 1997: "Incidence de la nature du sol et du cépage sur la maturation du raisin, à Saint emilion, en 1995", In: Colloque international "Les terroirs viticoles", 17–18 juillet 1996. Angers, 154–157.

Van Leeuwen, C., Seguin, G., 2006: "The Concept of Terroir in Viticulture", in: *Journal Wine Research*, 17, 1: 1–10.

Van Leeuwen, C., Gaudillere, J.P., Tregoat, O., 2001: "Evaluation du régime hydrique de la vigne à partir du rapport isotopique $^{12}C/^{13}C$", in: *Journal International Science Vigne Vin*, 4: 195–205.

Van Leeuwen, C., Tregoat, O., Chone, X., Jaeck, M.E., Rabusseau, S., Gaudillere, J.P., 2003: "Le suivi du régime hydrique de la vigne et son incidence sur la maturation du raisin", in: *Bull. O.I.V.* 867, 868: 367–379.

Van Leeuwen, C., Friant, P.,Choné, X., Tregoat, O., Koundouras, S., Dubourdieu, D., 2004: "Influence of Climate, Soil, and Cultivar on Terroir", in: *American Journal Enol Vitic,* 55, 3: 207–217.

Van Leeuwen, C., Tregoat, O., Choné, X., Bois, B., Pernet, D., Gaudillere, J.P., 2009: "Vine Water Status is a Key Factor in Grape Ripening and Vintage Quality for Red Bordeaux Wine. How Can it be Assessed for Vineyard Management Purposes?" in: *Journal International Science Vigne Vin,* 43, 3: 121–134.

Vaudour, E. 2002: "The Quality of Grapes in Relation to Geography: Notions of Terroir at Various Scales", in: *Journal Wine Research,* 13, 2: 117–141.

Vaudour, E., 2003: "Les Terroirs Viticoles", Dunod (Ed.). Définitions, caractérisation et protection, Paris, France.

Vavoulidou, E., Avramides, E.J., Dimirkou, A., Papadopoulos, P., 2006: "Influence of Different Cultivation Practices on the Properties of Volcanic Soils on Santorini Island", in: *Greece. Communications in Soil Science and Plant Analysis*, 37, 15–20: 2857–866.

Villeneuve, F., 1997: "Growing soil and vegetable quality. Example: the carrot./terroir et qualité des légumes. Exemple: la carotte. Infos (Paris)", 136:41–44.

Viscarra Rossel,. R.A., Cattle S.R., Ortega A., Fouad Y., 2009: "In situ Measurements of Soil Colour, Mineral Composition and Clay Content by Vis-NIR Spectroscopy", in: *Geoderma*, 150, 3–4: 253–66.

White, R. E., 2003: "*Soils for Fine Wines"* (New York: Oxford University Press).

White, R., Balachandra, L., Edis, R., Chen, D., 2007: "The Soil Component of Terroir", in: *Journal of International Science Vigne Vin,* 41, 1: 9–18.

Witbooi, E.H., Carey, V.A., Hoffman, J.E., Strever, A.E., 2008a: "The Relationship Between Soil Surface Colour and the Performance of *Vitis vinifera* L. Cv. Cabernet Sauvignon in

Stellenbosch Wine of Origin District. I. Vegetative Growth", 31st World Congress of Vine and Wine and the 6th General Assembly of the OIV, Verona, Italy.

Witbooi, E.H., Carey, V.A., Hoffman, J.E., Strever, A.E., 2008b: "The Relationship Between Soil Surface Colour and the Performance of *Vitis vinifera* L. Cv. Cabernet Sauvignon in Stellenbosch Wine of Origin District. II. Yield, Berry—and Wine Composition", 31st World Congress of Vine and Wine and the 6th General Assembly of the OIV, Verona, Italy.

Wong, M.T.F., Harper, R.J., 1999: "Use of on-Ground Gamma-Ray Spectrometry to Measure Plant-Available Potassium and Other Topsoil Attributes", in: *Australian Journal of Soil Research*, 37: 267–277.

Yadessa, A., Burkhardt, J., Denich, M., Woldemariam, T., Bekele, E., 2008: "Influence of Soil Properties on Cup Quality of Wild Arabica Coffee in Coffee Forest Ecosystem of SW Ethiopia", ASIC Conference Proceedings, Campinas 2008.

Zhang, J., 1966: "Relation Between Orange Quality and Geochemical Background of Soil", in: *Hunan Geology*.

Zhang, J., Davies, W.J., 1987: "Increased Synthesis of ABA in Partially Dehydrated Root Tips and ABA Transport from Roots to Leaves", in: *Journal of Experimental Botany*, 38: 2015–23.

Zhang, J., Davies, W.J., 1989: "Abscisic Acid Produced in Dehydrating Roots May Enable the Plant to Measure the Water Status, in: *Plant Cell and Environment*, 12: 73–81.

Zhou, H., Liu, J., 1997: "The determination of rare earth elements in plant foods by ICP-MS", in: *Atomic Spectroscopy*, 18, 6: 192–94.

Zhou, Q., Kang, S., Zhang, L., Li, F., 2007: "Comparison of APRI and Hydrus-2D Models to Simulate Soil Water Dynamics in a Vineyard Under Alternate Partial Root Zone Drip Irrigation", in: *Plant and Soil*, 291, 1–2: 211–23.

Zsófi, Z.S., Gál, L., Szilágyi, Z., Szűcs, E., Marschall, M., Nagy, Z., Bálo, B., 2009: "Use of Stomatal Conductance and Pre-dawn Water Potential to Classify Terroir for the Grape Variety Kékfrankos", in: *Australian Journal Grape and Wine Research*, 15, 1: 36–47.

Chapter 7
Polygenic Red Calcic Soils in Coastal Middle Palaeolithic Environments, Israel: Taxonomy and Pedosedimentary Reconstructions

Alexander Tsatskin

Abstract This paper reports the results of the ongoing multidisciplinary research of the Late Pleistocene palaeosol sequences in cemented aeolianitc in the Carmel coastal plain. The sequence at the Habonim quarry (Tsatskin et al. 2009) is proposed as type section for the Habonim pedocomplex, related to the Last Interglacial *sensu lato* at least. In other localities in the Carmel coast numerical dates obtained thus far substantially deviate from those at Habonim. Although lacking the degree of chronological resolution found at the type section, other palaeosols with Mousterian finds are also polygenetic and include several pedogeomorphic stages. This allows us to roughly correlate red calcic palaeosols with the Habonim pedocomplex. Macrorhizoliths embedded in such welded palaeosol sequences are shown to post-date the major phase of pedogenesis. The major palaeosol at the site near Atlit is taxonomically identified as polygenetic Hamra or Hamra-Husmas (Israeli classification), which developed under wetter than today climate, stronger desert dust deposition, carbonate leaching, and reddening. Before burial, the palaeosol likely developed in an unstable environment under accelerated sand accretion, plausible dust events, and incipient calcretization. Comparisons with surface sandy soils, though poorly preserved in the studied area, are attempted.

Keywords: Aeolianite · Hamra · Husmas · Late pleistocene · Carmel coast

7.1 Introduction

During the Quaternary, the Mediterranean coastal plain of Israel underwent cyclic changes related to fluctuation of climate and sea level. The environmental changes apparently affected two-sided migrations of anatomically modern humans and

A. Tsatskin (✉)
Zinman Institute of Archaeology, University of Haifa, 31905 Haifa, Israel
e-mail: tsatskin@research.haifa.ac.il

S. Kapur and S. Erşahin (eds.), *Soil Security for Ecosystem Management*, SpringerBriefs in Environment, Security, Development and Peace 8,
DOI: 10.1007/978-3-319-00699-4_7,

Neanderthals at the crossroads of the eastern Mediterranean. By the end of the Pleistocene, $\sim$15 ka BP, new forms of settled life appeared in southern Levant which gradually paved the way for intentional use of soils for food production (Bar-Yosef and Belfer-Cohen 1989). The suitability of Mediterranean soils for ancient agriculture was apparently one of the key prerequisites for Neolithic economy. However, the history of the development of surface Mediterranean soils is as yet difficult to establish. As Yaalon (1997) pointed out, soils of Mediterranean climates, in particular terra rossa, are extremely complex open systems with persistent addition of Saharan dust and a long history of anthropogenic impact. Hence, archaeology and numerical dating techniques are necessary to obtain a better understanding of how Earth's critical zone functions over time (Holliday 2004).

Alternately, soils are one of the best archives of environmental change records, in particular in loess/soil sequences of the world. However, the significance of soils and palaeosols in Mediterranean coastal sandy areas, termed Hamra (Dan et al. 2007; Singer 2007), for reconstructions of palaeoclimate is still ambiguous. The western part of the Israeli coastal plain is occupied by sand dunes mainly composed of quartz which is delivered to the shore from the Nile delta by the eastern Mediterranean currents (Golik 2000). Longitudinal, north–south ridges of calcareous aeolianite (termed kurkar), cemented in the course of calcareous digenesis of marine sand, form key geomorphic surfaces in the plain (Tsoar 2000). In their recent review paper, Sivan and Porat (2004) conclude that kurkar and Hamra of the coastal plain formed at different time intervals throughout the Late Pleistocene, and that their formation is not related to any particular climatic event. This is controverted by Gvirtzman and Wieder (2001) and others.

In this paper, we will show that pedosedimentary sequences in aeolianite in Mediterranean areas seem to provide important information on Quaternary environmental change. We focus on the Late Pleistocene soil development and landscape evolution in the Mediterranean coastal plain adjacent to Mount Carmel, called the Carmel coast (Fig. 7.1). In this part of the plain, abundant exposures of kurkar, Hamra, and karstic phenomena all are found in road cuts along the Haifa–Tel Aviv highway and in quarries (Fig. 7.2). In addition to reviewing earlier studies, this paper reports a new study on soils from the site near Atlit, termed *Atlit Junction Bridge* (AJU) in Frechen et al. (2004). Hamra soil in Israeli nomenclature is an equivalent for Rhodoxeralf (Soil Taxonomy 1999) or Chromic Luvisol (WRB 1998). Numerous Mousterian prehistoric artefacts were found in Hamra palaeosols of the Carmel coast (Ronen 1977). In contrast to surface Hamra-Grumusol (Vertisol) catena from Sharon (Dan et al. 1969), Hamra palaeosols in the Carmel coastal plain are strongly calcareous and have indurated calcarenite in substrate. In Israeli classification (Dan and Koyumdjiski 1979), calcareous variants of Hamra are called Husmas. Our goal was to show the palaeo-environmental and stratigraphic significance of palaeosols in the Carmel area via refinement of palaeosol

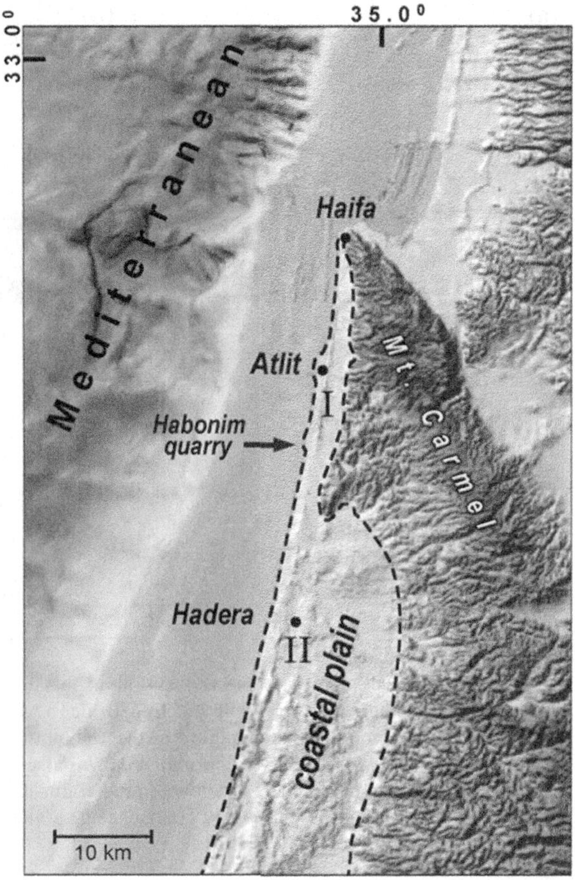

Fig. 7.1 Location map of coastal plain of northern Israel from Haifa to Hadera (*I*—Carmel coast, *II*—Sharon coast) prepared on the basis of the digital shaded relief map of Israel 1:5,00,000 by Hall and Calvo. *Source* This author

taxonomy and differentiation between diagenetic and pedogenic carbonate. The age of palaeosols in the AJU site was obtained by luminescence (Frechen et al. 2004), and that of surface soils by the ^{14}C method.

7.1.1 Methods

We studied surface soils and palaeosols in and around prehistoric archaeological sites earlier described along the Carmel coast (Farrand and Ronen 1974). Buried in kurkar palaeosols were studied in road cuts of the Haifa–Tel Aviv highway. Field studies focused on catenary relationships (toposequences) and detailed soil morphology that included description of texture, Munsell colour notation, structure, abundance and type of neoformations, etc. Total carbonates were measured by volumetric calcimetry. Particle size analysis was carried out by a hydrometer

Fig. 7.2 Field view of **a** poorly preserved natural maquis upon kurkar ridge near Atlit; **b** road cut of leeward side of kurkar at the Atlit Junction Bridge (AJU), 9 km Haifa–Tel Aviv highway; **c** exposure of the Habonim pedocomplex in the Habonim quarry (location shown in Fig. 7.1) showing *dark-coloured* Vertisol on *top* and *reddish* palaeosol in *lower part*; **d** macrorhizolith from the *middle part* of *red* palaeosol, shown in Fig. 7.2b, at AJU near Atlit; **e** karstic cavity filled with *reddish* soil material of supposedly Holocene age at the windward side of kurkar ridge near Atlit ~30 ka BP. *Source* This author

method after dispersion in calgon. Soil micromorphology was performed on 24 × 36 thin sections, which were prepared after polyester resin #442 plus acetone impregnation of samples under vacuum. Thin sections descriptions follow Stoops (2003). Digital colour indices a* (redness), b* (yellowness), and L* (lightness) were measured by spectrocolorimeter Specbos 4000 (JETI, Jena). Low-field mass susceptibility at 0.47 (χ_{lf}) and 4.7 frequency were measured by Bartington MS2 magnetic susceptibility meter (Oxford, UK). The difference between the two, frequency-dependent susceptibility (χ_{fd}), is a measure of very small *superparamagnetic* (SP) grains (Evans and Heller 2003).

The ^{14}C dates were obtained by *accelerator mass spectrometry* (AMS) in Poznan Radiocarbon Laboratory, Poland. The samples were treated chemically according to the standard AAA (acid-alkali-acid) procedure. After chemical pretreatment, the samples were combusted in sealed quartz tubes (with CuO and Ag), and the CO_2 produced was purified and graphitized by reduction with H_2, using Fe powder as a catalyst (Czernik and Goslar 2001).

7.2 Late Pleistocene Pedostratigraphy

7.2.1 Israeli Coastal Plain

The type sections of the Late Quaternary stratigraphic sequence of the Israeli coastal plain were described and analyzed (Gvirtzman et al. 1998; Gvirtzman and Wieder 2001) in the central part of the coastal plain, termed Sharon plain (Fig. 7.1). A thick kurkar at the base of the Late Pleistocene sequence is dated by luminescence to ~53,000 BP and is overlain by the Netanya palaeosol sequence, that includes from below upward lower Hamra, middle Grumusol (Vertisol), and uppermost Hamra all covered by a thin kurkar, ~5 ka BP.

According to Porat et al. (2004), sand accretion occurred on the Israeli coast rapidly during the formation of kurkar. However, during Hamra formation and maturation, deposition of fine dust from distant Saharan sources may have accelerated (Danin and Yaalon 1982; Yaalon 1997).

7.2.2 Carmel Coast

In contrast, the Carmel coast provides a different pattern of pedosedimentary evolution. The Carmel coast is the narrowest part of the Israeli coastal plain, with the one eastern onshore kurkar ridge up to 18–20 m above sea level (Fig. 7.2a, b). The western kurkar ridges between Atlit and Haifa are now submerged (Sivan and Porat 2004).

Recently, sedimentological and geochronological studies have been carried out on the Carmel coast on a systematic basis (Neber 2002; Frechen et al. 2004). Chronological data showed the existence of several periods of dunes accretion intermittent with soil formation. However, those periods were inferred not in a complete stratigraphic sequence but compiled from different sites. The only exception is the site at the Habonim quarry (Fig. 7.2c; for location see Fig. 7.1). Within one sequence, a lower reddish palaeosol dated to >100 ka BP (Ronen et al. 1999; Tstaskin and Ronen 1999) grades upward into a Vertisol buried by a young kurkar, ~30 ka BP. Thus, the palaeosol at Habonim is significantly older than that of the Netanya sequence in the Sharon area. Taking into account also that in the Carmel coast Middle Palaeolithic tools are occasionally found inside the palaeosol, and marine deposits related to *marine isotopes stage* (MIS) 5e are intercalated with kurkar near Haifa (Neber 2002), one may question the applicability of Gvirtzman and Wider's (2001) scheme for the Carmel coastal plain.

7.2.3 Habonim Pedocomplex

We propose to use the site at Habonim quarry as a type section in the Carmel coast in which a thick palaeosol sequence formed during the ~120–45 ka BP interval (MIS 5 and 4) and is buried by terminal Pleistocene kurkar. This sequence will be referred to as the Habonim pedocomplex (Fig. 7.2c). The Habonim palaeosol sequence is 4.5 m thick, occupies an interdune depression ~100 m long, and is covered by aeolianite dated to ~30 ka BP. The sequence is composed of several contrasting and partially welded palaeosols which are not separated by non-soil sediments (Fig. 7.3).

The studies of thin sections, environmental magnetism, clay mineralogy, and the Mössbauer effect have all allowed us to propose a refined Late Pleistocene evolution of soil and sedimentary processes (Tsatskin et al. 2009). As shown in Fig. 7.3, the numerical dates are in good chronological order, apart from those in the lower kurkar. According to the dates, Hamra soil at the bottom of the Habonim sequence may be related to MIS 5 *sensu lato*. The sequence shows three peaks of pedogenic carbonates at 1.5, 2.4, and 4.7 m depth (Fig. 7.3), decrease of magnetic susceptibility from magnetically enhanced Hamra to the overlying Vertisol, and an opposite trend for clay fraction amount. In Hamra, ferromagnetic and

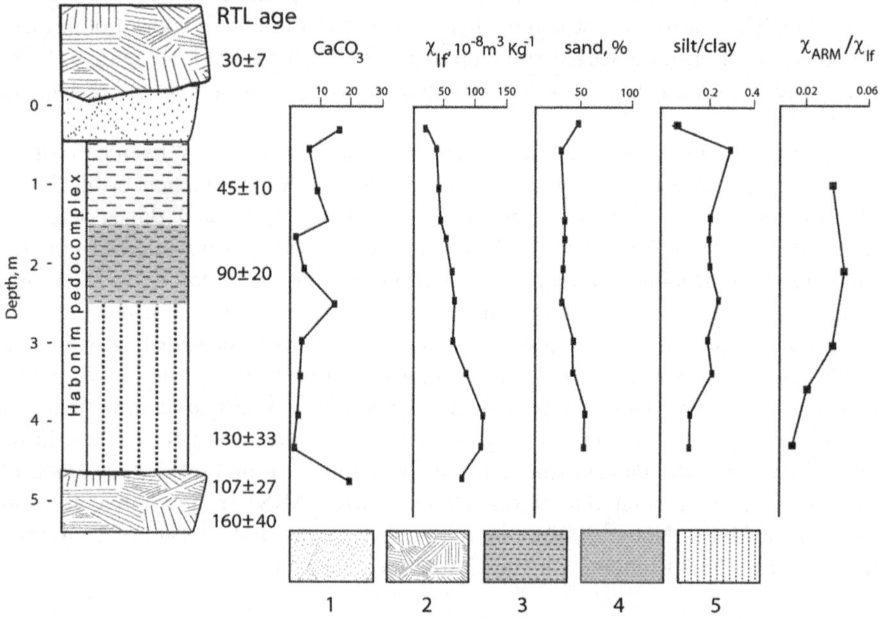

Fig. 7.3 Some properties and numerical ages of the Habonim pedocomplex at the type section related to MIS 5 through 4 (3); *1*–sand; *2*–kurkar; *3*–vertic soil with gley features and disturbances; *4*–dark vertic soil with strong slickensides; *5*–polygenic Hamra (typic Rhodoxeralf). *Source* Reproduced with permission of the Journal of Mountain Science

superparamagnetic minerals accumulated, whilst in younger, poorly drained Vertisol ferromagnetic minerals underwent reductive dissolution. Significantly, the input of fines during the prolonged pedogenesis was not uniform, as shown in Fig. 7.3 by both a slight upward silt/clay ratio increase and a coarsening of wind-blown magnetic particles. The latter is measured by magnetic grain size parameters such as the ratio of anhysteretic remnant magnetization susceptibility (χ_{ARM}) to χ_{lf} (Evans and Heller 2003), which increases upward.

7.3 Kurkar Ridge from Haifa to Atlit

7.3.1 Surface Sandy Soils

Near Atlit the patches of natural Mediterranean maquis vegetation grow on sandy soils upon the cemented aeolianite (Fig. 7.2a). Sand mantle has its maximal thickness on the windward side of the ridge. The surface has an undulating topography with higher areas alternating with slight depressions in which soil cover is maximal. On summits, thin soils and kurkar outcrop. Variable thickness of the soil cover may indicate transport of soil materials from the upper to the lower areas. The dunes are covered by sparse trees and shrubs such as Olive (*Olea europea*) and hairy bow (*Callocotome villosa*), as well as grain grasses such as goddess gum (*Pistacia lintiscus*), shaggy daphne (*Thylaea Hirsuta*), and others (see Flora of Israel Online). The area is disturbed by human activity, which is probably why carob, oak, and other Mediterranean tree species (Kutiel 2001) do not grow here.

Tracing the surface soil along railroad cuts allowed us to recognize occasional small-scale epikarstic forms. The karstic cavity (Fig. 7.2e) is filled with red sandy loam which does not form a horizontal horizon in a surface soil. A similar ponor-like feature is present in the road cut on the leeside of the ridge (Fig. 7.4).

A representative soil profile 3–10 was dug on the summit of the kurkar ridge where soil is covered by a sand sheet described below as unit I.

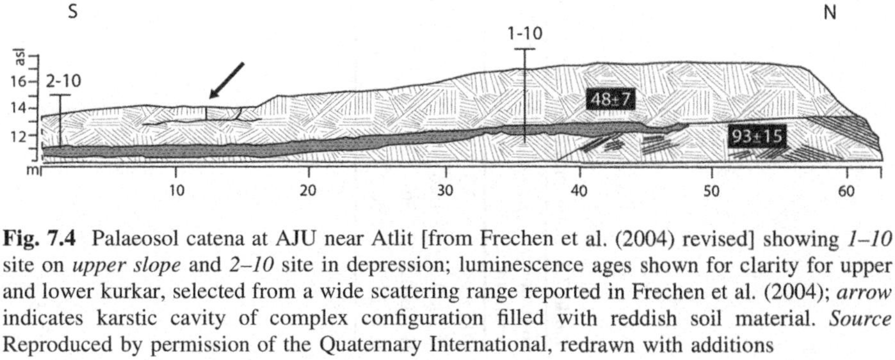

Fig. 7.4 Palaeosol catena at AJU near Atlit [from Frechen et al. (2004) revised] showing *1–10* site on *upper slope* and *2–10* site in depression; luminescence ages shown for clarity for upper and lower kurkar, selected from a wide scattering range reported in Frechen et al. (2004); *arrow* indicates karstic cavity of complex configuration filled with reddish soil material. *Source* Reproduced by permission of the Quaternary International, redrawn with additions

Table 7.1 Some properties and AMS ^{14}C dating of surface soil. *Source* This author

Horizon/ depth, m	$\chi_{lf,\ 10\text{-}8\ mkg}^{3\quad -1}$	χ fd, %	L*	a*	$CaCO_3$ %	OM %	Sand %	Silt %	Clay %	AMS ^{14}C BP
I/0.1	3.3	nd	53.08	6.39	41.1	0.32	87.7	1.9	10.4	nd
II Ak/0.3	18.6	8.0	40.86	9.7	8.9	0.93	75.7	7.9	16.4	Poz-38218: 85 ± 35
II AB/0.5	36.3	7.1	36.37	11.38	0.8	0.65	67.7	7.9	24.4	Poz-38241: 400 ± 30
II B/1.0	46.2	8.7	37.61	13.07	0.8	0.55	67.7	7.9	24.4	Poz-38238: 850 ± 30

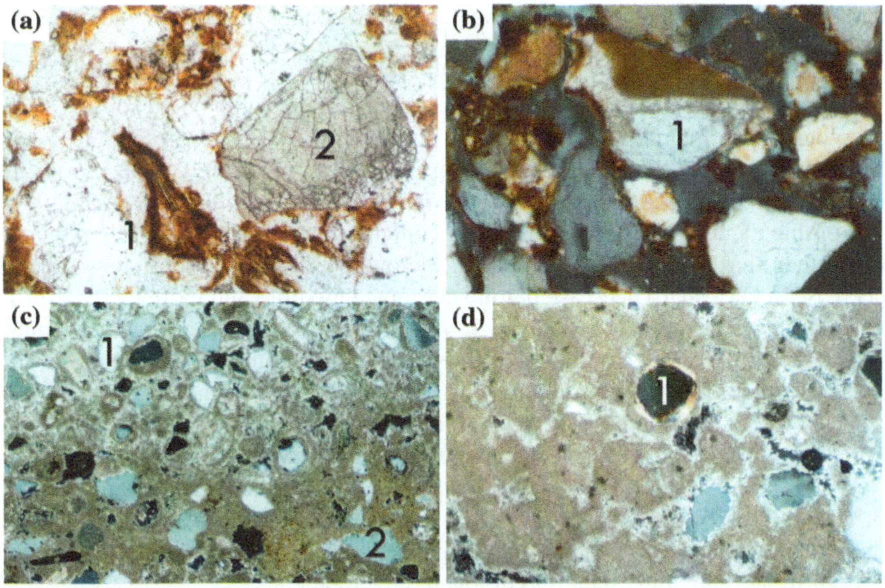

Fig. 7.5 Photomicrographs of some micromorphological features—from surface soil, *plane polarized light* (PPL), *crossed polarized light* (XPL); **a** at 1 m depth, strongly decayed remnant of rootlet (*1*) and calcite shell (*2*) with signs of recrystallization on lower surface, length of frame 1.01 mm, PPL; **b** an allochem bound with quartz sand grain by sparite (*1*) in open porphyric related distribution pattern, length of frame 1.01 mm, XPL; from youngest kurkar in a surface soil substrate; **c** juxtaposition of sparitic matrix with embedded sand-sized grains with isopachous rims (*1*) and dense micritc matrix with less abundant coating-free grains (*2*), length of frame 2.54 mm, XPL; **d** micritic matrix with 10 % embedded quartz in rhizolith, note abundant dissolution cracks and planes with recrystallization of calcite into microsparite, as well as a specific clay-coated rounded sand grain (*1*) whose origin is explained in the text, length of frame 2.54 mm, XPL. *Source* This author

Unit I. Mantle of loose sand, 0.3–0.4 m thick, 10YR 6/4 light yellowish-brown well sorted, abounds with marine shell detritus, in other patches more stabilized; lower contact is sharp, with irregular narrow subvertical fissures to ~0.5 m deep.

Unit II includes a soil profile that consists of the following horizons. A_k horizon, 0.2–0.3 m thick, 7.5YR 4/6 strong brown sandy loam, heterogeneous due to fine mottles of burrowing insects; strong effervescence, loose, some peds are compacted, porous, abundant rootlets, rare small soft $CaCO_3$ nodules; gradual transition.

- $AB_{(k)}$ horizon, 0.2–0.6 m thick, 5YR 4/6 yellowish-red sandy loam, finely mottled due to bioturbation, firmer than above, weak effervescence in spots; sharp transition.
- $B_{(k)}$ horizon, 0.6–1.2 m thick, 5YR 4/6 yellowish-red sandy loam, firm, structure poorly preserved, except for occasional 3–5 cm peds, with no lining on walls,

finely mottled, fine porosity, weak effervescence in spots; underlain by cemented kurkar.

Analytical measurements are in good agreement with field observations showing that a mantle of loose sand with 87.7 % sand, 41.1 % $CaCO_3$, and a negligible χ_{lf} value of 3.3×10^{-8} m^3 kg^{-1} (Table 7.1) strongly differs from the underlying soil. The A_k horizon shows 0.93 % *organic matter* (OM), maximal in the profile, and appreciable carbonate content (8.9 %), the latter significantly lower than that in the cover sand sheet. In contrast, lower horizons show gradual decrease in OM, sharp decrease in total carbonate, twofold and more increase in χ_{lf} values. The latter coincides with high values of a* redness index, which equals 13.07 units in the lowermost B horizon at 1.0 m depth. In thin sections soil material has a loose structure up to 40 % pore space with abundant channels, and consists of diffuse crumbly peds, with isotropic organic-clay mass lining sand grains. Abundant organic residues, some strongly decayed, are found throughout (Fig. 7.5a). Small, ~0.5 mm sparitic nodules that bind quartz grains and remnants of a dissolving bioclast occur (Fig. 7.5b). Recrystallized calcareous bioclasts are normally embedded in isotropic calcite-free matrix. The soil is characterized overall by an immature stage of carbonate development.

Although the surface sandy soil may be defined as Regosol, the lower $B_{(k)}$ horizon of strong red colour, as confirmed by spectral indices (see above), may have constituted the relict of earlier stronger Hamra. However, the AMS ^{14}C dates, spanning a ~1,000 year interval (Table 7.1) seemingly rule out the latter possibility. On the other hand, the whole profile shows signs of strong biological burrowing by small insects (ants, nematodes) and penetration by deep rootlets, which may have persistently rejuvenated the ^{14}C activity. As emphasized by Paton et al. (1995), bioturbation in sandy soil materials is a powerful near-surface process in coastal settings worldwide.

7.3.2 Red Calcic Palaeosols with Mousterian Artefacts

The Mousterian type artefacts are found here in the upper part of the palaeosol, as in other Carmel localities. Figure 7.4 shows a palaeosol catena from the 1–10 site on the upper slope to the 2–10 site in an interdune depression in the AJU site. The samples from upper and lower kurkar were dated by luminescence and results showed surprisingly large scattering (Frechen et al. 2004). Because of this we selected two dates that were the most realistic in our opinion to be shown on Fig. 7.4. The dates provide a decisive criterion to assign to the palaeosol at the AJU site a stratigraphic position younger than the Habonim pedocomplex (Frechen et al. 2004). Let us look at the properties of the 1–10 palaeosol more closely. The 1–10 site has the following morphology (depth from the upper contact).

AB_k1 horizon, 0–0.5 m, 5YR 5/6 yellow red sandy loam, compacted, porous, strongly calcified in the form of impregnations along pores with diffuse boundary

Table 7.2 Some properties of red calcic palaeosol near Atlit (90th km Haifa–Tel Aviv highway). *Source* This author

Site/ depth*, m	χ_{lf}, $10^{-8}\,m^3kg^{-1}$	χ_{fd}, %	L*	a*	b*	$CaCO_3$, %	Sand, %	Silt, %	Clay, %
1/0.2	64.5	9.2	48.88	12.53	19.9	10.5	68.2	2	29.8
1/0.4	79.8	9.1	43.45	14.15	20.66	6	68.2	2	29.8
1/0.5	88.0	9.1	43.02	14.46	21.18	5.2	64.2	2	33.8
1/0.8	93.5	8.8	43.03	13.68	20.47	2.4	56.2	6	37.8
–"–burrow red	nd		40.41	14.54	19.34	nd	nd	nd	nd
–"–burrow calcareous	nd		42.71	14.11	19.9	nd	nd	nd	nd
1/1.1	96.3	8.8	47.85	12.02	20.06	11.3	nd	nd	nd
1/1.2	93.4	9.0	44.99	12.22	19.44	10	52.2	6	41.8
2/02	30.7	7.8	46.21	9.79	20.70	nd	nd	nd	nd
2/0.4	39.1	6.6	44.27	8.15	17.44	nd	nd	nd	nd

Depth from the upper contact of palaeosol; nd—not determined and hard small $CaCO_3$ concretions, as well as rhizoliths ranging from ~1 cm in diameter to ~4 cm thick (Fig. 7.2d); gradual transition.

AB_k2 horizon, 0.5–0.9 m, 5YR 4/6 yellow red loam/sandy clay loam, strongly compacted, some large wedge-shaped peds; strongly calcified with mycelium forms prevailing, in contrast to upper horizon, though carbonate concretions 3–5 cm in size are scattered throughout, at 0.7 m depth below the upper contact, some 2.5YR 4/8 red decalcified burrows (krotovinas) ~6 cm thick are found, gradual transition.

BC_k horizon, 0.9–1.2 m, 7.5YR 4/4 brown loam, firm, heterogeneous due to large calcrete-like concretions, gradually turning into carbonate crust-like horizon overlying indurated aeolianite (kurkar).

The palaeosol (Table 7.2) consists of upper sandy loam with 68.2 % sand and 29.8 % clay in the uppermost horizon at 0.2 m depth, and a loam with only 52.2 % sand and 41.8 % clay at 1.1 m depth. The uppermost horizon is enriched with carbonates (10.5 %), whose amount drops to 2.4 % in the middle of the profile and increases again at the bottom where calcrete-like nodules abound. Redness index a* has high values in the middle of the profile and reaches maximum in a rodent burrow filled with non-calcareous red loam (14.54 units). Magnetic enhancement with χ values, varying in the range 80–95 $\times$ 10^{-8} m^3 kg^{-1}, characterizes both the middle and the lower part of the profile, substantially dropping to 64.5 $\times$ 10^{-8} m^3 kg^{-1} at the uppermost sand-rich calcareous horizon. Although Munsell readings are similar in a surface soil and a palaeosol, digital colour indices indicate that in surface soil the reddest B horizon is slightly less red and substantially darker than in the red palaeosol, with ~13 a* vs. 14.54a* and 40.41 L* vs. 37.61 L*, respectively (Tables 7.1 and 7.2). Note that white colour has L* of 100 units.

In contrast, the 2–10 site in the depression (Fig. 7.4) is characterized by a palaeosol whose upper horizons show low a* colour index and depleted magnetic susceptibility (Table 7.2). This is explained by intense dissolution of ferromagnetic minerals under reduced conditions in Vertisols (Tsatskin et al. 2009).

In sum, field observations along with analytical properties are not conclusive enough for taxonomic identification of red palaeosols in the 1–10 section because of the abundance of carbonate. According to accepted criteria (Dan and Koyumdjiski 1979; Gvirtzman and Wieder 2001), genetically similar soils with strong reddening of the fine mass are classified Hamra, if leached from carbonates, and Husmas, if an appreciable amount of $CaCO_3$ is contained in the fine mass. In addition, the palaeosols described here are found above an indurated aeolianite, as is the case in many localities in the coastal plain. As suggested by Yaalon (1967), aeolianite is rather a substrate for red palaeosols and not their parent material. In order to elucidate these questions we focus on soil micromorpholgy with special attention to secondary carbonate.

7.3.3 Secondary Carbonates

In the Carmel coast, Neber (2002) described several types of aeolianites on the grounds of sedimentary bedding, type and extent of cementation, development of rhizoliths, and epikarstic features of cavernous weathering. Within the study area, Neber differentiated between massive, nodular kurkar (Fig. 7.2b, top) and kurkar with cross-bedded foresets in a leeward position (Fig. 7.2b, bottom). He also attempted to distinguish between phreatic and vadose meteoric type of digenesis. However, calcareous cement in aeolianites is strongly variable at a local level (McLaren and Gardner 2004), depending on texture, plant cover, exposure to sea spray, etc.

In thin sections a massive calcarenite is almost devoid of marine bioclasts let alone tiny fragments of foraminifera, algae, and shells. The fabric is constituted by well-sorted sand grains of quartz coated with isopachous micritic rims which are embedded in either micritic or more often microsparitic cement (Fig. 7.5c). Generally, grains with isopachous rims are taken as a diagnostic feature for a meteoric vadose diagenesis. In addition, some denser, rounded patches show later diagenetic transformations (Fig. 7.5c, bottom) recognizable as dissolution of isopachous rims and formation of clean microsparitic calcite overgrowths in cracks and vughs (Fig. 7.5d). We diagnose such patches as evidence for root casts (rhizocretions) apparently from vegetation supported by a dune. The formation of rhizocretions results from the precipitation of carbonate from saturated solutions that migrate downward. This means that surface karstic processes may have occurred at the time of later diagenesis of kurkar.

7.3.4 Pedogenic Carbonate

Studies in the micromorphology of secondary carbonate in Mediterranean soils by Wieder and Yaalon (1974, 1982) showed that (1) in clayey soil materials carbonate normally precipitates in micrite form, and (2) nodules containing skeletal grains and diffuse boundaries (orthic) formed in situ while nodules free of skeletal grains (allothic) were rather transported by colluviation. Recent studies of larger soil nodules also demonstrate the importance of distinct stages of secondary carbonate development related to environmental change (Tsatskin et al. 2009).

The 1–10 palaeosols in AJU contains (1) small-sized, poorly consolidated carbonate; (2) macrorhizoliths similar to those described by Khadkikar et al. (2000) and Alonso-Zarza et al. (2008) either in vertical or inclined configuration (Fig. 7.2d; and (3) large crust-like concretions at the base of the soil profile. The macrorhizoliths are tentatively attributed to a development stage no higher than stage II (Gile et al. 1966).

Thin sections confirm the analytical data that soil carbonates peak both in the upper and lower parts of the profile, while the middle part is practically leached

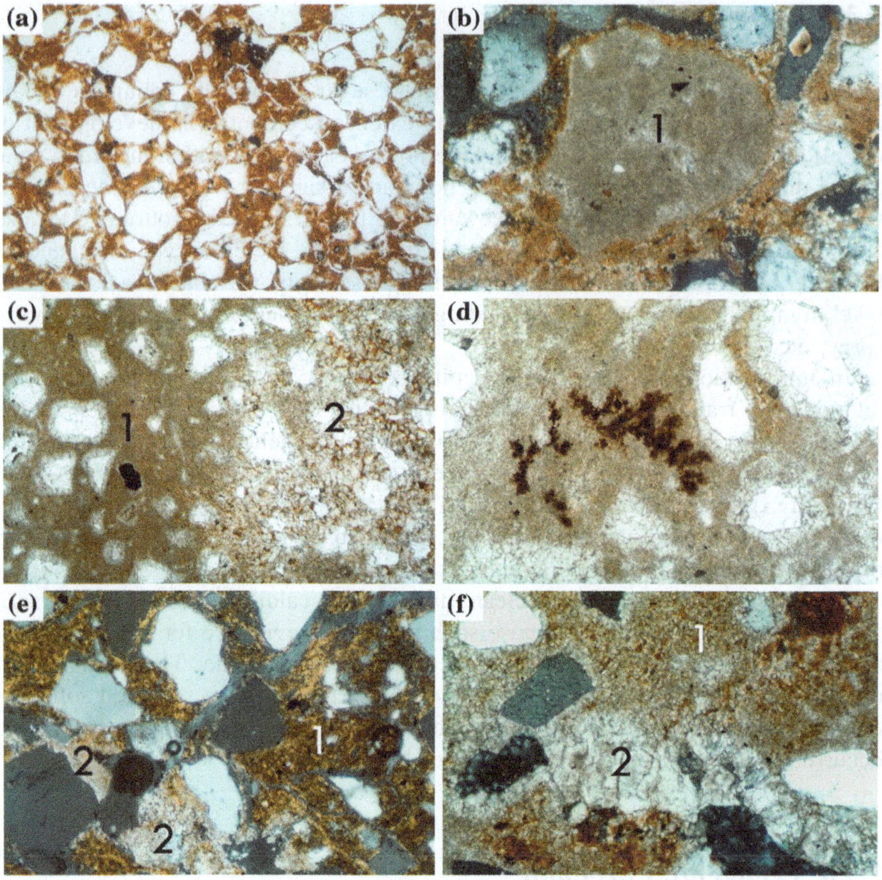

Fig. 7.6 Photomicrographs of the Habonim pedocomplex at the 1–10 site near Atlit; plane polarized light (PPL), crossed polarized light (XPL); **a** dense porphiric distribution pattern of well sorted sand-sized quartz unevenly distributed in *brown red* groundmass with some *curved planes* in the major decalcified phase of red palaeosol, at 0.7 m depth, length of frame 2.54 mm, PPL; **b** allotic micritic nodule ~0.7 mm in size coated (*1*) with illuviated *dusty brown* (aged) clay from the red calcic palaeosol of the final stage of pedocomplex development at 0.3 m depth, length of frame 1.01 mm, XPL; **c** juxtaposition in macrorhizolith of *dense brown*-stained micrite in which sand grains are coated with microsparite recrystallized in a perpendicular to grain surface fashion (*1*) grades to strongly reworked and recrystallized microsparite groundmass (*2*) enriched with strongly fragmented clay coatings; note gradual vanish of sparite-coated quartz, length of frame 2.54 mm, PPL; **d** diffuse Fe–Mn impregnations in the micritic-microsparitic *brown* stained groundmass of macrorhizolith, length of frame 1.01 mm, PPL; **e** strong strial b-fabric (*1*) of the lower horizon of pedocomplex with discontinuous (probably disrupted) micrite coating (*2*) around pores at 1.2 m depth, length of frame 1.01 mm, XPL; **f** complex calcite nodule at 1.2 m depth in which strongly fragmented aged clay coatings are embedded along with microsparite coated quartz in dense micrite matrix (*1*), note the lining (*2*) composed of clean equant sparite grains ~90 μm, length of frame 1.01 mm, XPL; more explanation in text. *Source* This author

(Table 7.2). In the decalcified middle part of the profile, where a* redness is maximal and reaches ~14.5 units, the microfabric shows porphyric related distribution with subangular to subrounded sand-size quartz grains, averaged 0.1 mm in size, embedded in a red-brown groundmass with striated-speckled b-fabric (Fig. 7.6a). Abundant curved planes delineate incomplete spheroidal peds. Significantly, these features, albeit highly developed, were described by Kapur et al. (1997) in surface Vertisols in the eastern Mediterranean. In contrast, the upper part of the profile (Fig. 7.6b) containing ~10 % carbonate shows allothic nodules ~0.7–0.8 mm in size coated with brown–red clay. Allothic nodules are an indication of erosion/deposition apparently at the final stage of pedocomplex development. Dusty clay coatings surrounding skeletal grains, as well as juxtaposition of clay illuviation features upon allothic nodules, indicate that clay infiltration post-dates calcification. In other words, upper calcareous palaeosol was forming upon mature vertic like Hamra after a period of time elapsed. Unfortunately the resolution of numerical dates (Frechen et al. 2004) is too low to assess the duration of the gap between upper and lower soils, which eventually welded.

Macrorhizoliths show an extremely complex microfabric, in fact, consisting of several superimposed types, e.g. (a) very dense micrite nodule with sand grains coated with clear microsparite rim, reminiscent of later diagenesis in rhizoliths found in kurkar (see Fig. 7.5b, c) microsparite/sparite mass incorporating tiny fragmented remnants of red-brown aged clay coatings (Fig. 7.6c). Within the macrorhizolith we observed several nucleation centres that eventually welded. The (b) type is evidence for partial dissolution of micrite and destruction of a rhizolith seemingly due to the onset of environmental change of wetter conditions, deeper percolation of clay suspensions, and stronger hydromorphism, revealed by occasional diffuse iron-manganese impregnations (Fig. 7.6d). The presence of a fragment with fabric typical for kurkar allows us to suggest that the microrhizolith studied incorporates a remnant inherited from an episode of kurkar accretion which post-dates the major soil phase and predates the final soil phase.

The lowermost horizon of the 1–10 profile in thin sections is represented as a juxtaposition of (a) strongly strial b-fabric of clayey mass and scattered thick distorted micritic coatings probably due to shearing (Fig. 7.6b, e) micritic-sparitic nodules of complex morphology. Figure 7.6f demonstrates juxtaposition of various calcite forms within a complex nodule, which suggests a sequence of developmental stages related to carbonate precipitation, its partial dissolution, and finally formation of an infilling with clean equant sparite.

7.4 Discussion

7.4.1 Calcareous Features Confirm Palaeosol Polygenesis

The strongly developed Habonim pedocomplex in its type section, composed of two/three contrasting accretionary palaeosols, likely developed under conditions of persistent aeolian-colluvial deposition during the ~120–45 ka time interval. In contrast, palaeosols at the site near Atlit are strongly welded and calcareous, and their numerical dates do not provide evidence for the duration of pedogenesis comparable with the Habonim pedocomplex. However, despite this, we propose to correlate the AJU welded palaeosol sequence with the Habonim pedocomplex.

This correlation is derived primarily from the complex polygenetic nature of a palaeosol sequence near Atlit, despite its shallowness. From the data on palaeosols elsewhere it is known that preservation of thick pedocomplexes that record discrete soil episodes is dependent on geomorphic conditions and on dust deposition rates (Fedoroff et al. 2010). It is not surprising then to find better "developed" pedocomplexes, e.g. on lower slopes and shallow depressions, as at Habonim (Tsatskin et al. 2009), which contrasts with palaeosol development near Atlit.

Juxtaposition of various microfabric types in macrorhizoliths that abound in the middle part of the pedocomplex (not on top) provides convincing evidence of their post-pedogenic origin in relation to the major phase of Hamra formation. They probably started to form during the gap in pedogenesis which marked the phase of accelerated sand accretion and continued into the next pedogenic episode. There were clear differences between those soil episodes in the intensity of carbonate leaching such that the younger episode was strong calcification whilst the major red palaeosol was strongly, yet not completely leached from carbonate. Recall that the major red palaeosol contains some stained micitic coatings in pores, occasionally distorted by shear stress, which by itself may be a sign of polygenesis of "lower hierarchy".

7.4.2 Palaeosol Taxonomy

Hence, the welded Habonim pedocomplex at the site near Atlit allows us to provide records of at least two discrete pedo-episodes—a major one and a final one. Between two episodes of pedogenesis a substantial period of time may have elapsed. The major palaeosol can be defined as Hamra (typic Rhodoxeralf) or Hamra-Husmas (calcic Rhodoxeralf) on the grounds that abundant carbonate in a major Hamra palaeosol was found to be unrelated to this pedogenic episode. In contrast, the final episode of soil formation is a less developed calcic red palaeosol, whose analogues we attempted to find among surface soils of the Israeli coastal plain.

Taxonomic identification of surface soils at the AJU locality turned out to be contentious due to their general similarity with poorly developed sandy Regosols, which is supported by young ^{14}C AMS. However, at ~1 m depth, Regosols show a developed soil horizon with a* redness ~13 units that is quantitatively similar to strongly red palaeosols in the 1–10 site (see Table 7.2), carbonate leaching, and feeble development of few calcite nodules. In contrast, assuming the post-pedogenic origin of strong calcareous concretions in an ancient red palaeosol, the latter should be viewed as incomparably more advanced stages of red soils development than preserved surface Regosols. In addition, taking micromorphological features as the key criteria, we see very advanced stages of red and yellow clay illuviation, their strong, intimate incorporation of ageing clay coatings in the groundmass, i.e. those criteria thoroughly elaborated by Fedoroff (1997) to define the mature stage of red Mediterranean soils. We have to remember that under neutral pH of red soils, clay coatings, albeit accumulating in B horizons, normally disintegrate in soil with high bioactivity, neutral pH, and possible $CaCO_3$ impact, undergo ageing, and are eventually incorporated into the matrix providing e.g. strial b-fabric (Stoops 2003).

7.4.3 Pedosedimentary Stages

In specific geological settings, red Mediterranean soils are increasingly understood as pedosedimentary sequences (Priori et al. 2008), particularly in coastal environments (Gvirtzman and Wieder 2001; Tsatskin and Ronen 1999). A new site at AJU clearly shows that during the *major* phase of Hamra formation the land surface seems to have aggraded slowly through dust deposition apparently from expanding/shrinking desert areas in the southern Levant in the earlier part of the Late Pleistocene and eventual dust imbedding in the soil profile (Yaalon 1997; Frumkin and Stein 2004). Since this palaeo-Hamra, as shown below, is by far much more strongly developed than the surface soil, the environment of soil formation was of stronger than present-day wetness and of denser vegetation. We also assume that this prolonged period was not completely stable, thus allowing for fluctuations in the groundwater table which created a sort of vertic behaviour in the Hamra soil.

Significantly, at the *final* stage of pedogenesis surface stability was likely disrupted, primarily due to coastal sands that started encroaching upon the surface of Hamra. We believe than no direct analogues from modern coastal settings may be selected for final pedogenic episodes roughly dated to MIS 4 or maybe later. On the basis of specific clay coated grains in thin sections, we follow Fedoroff et al. (2010), who also interpret them as formed via the interaction of fine dust and sand grains during rainstorms. Hence, environmental deterioration and probably dryness occurred, when, although reddening and illuviation ensued in soil, deposition of aeolian and colluvial sand increased, while macrorhizoliths and overall calcretization accelerated.

7.5 Conclusions

The new palaeopedological data from a new site near Atlit in the Carmel coast allows us to tentatively correlate a palimpsest of a shallow palaeosol sequence here with the highly resolved thick Habonim pedocomplex, notwithstanding chronological discrepancies existing thus far. Such interpretation has become possible through the refinement of palaeosol taxonomy and the establishment of palaeopedological stages. Such an integrative style of research seems critical for the better understanding of both present landscape processes and for the more rigorous applicability of palaeosols in coastal stratigraphy.

The resolution of fine changeability within the broad time interval spanned since the MIS 5 through 4 or 3 is not high in AJU, in contrast to Habonim. This is the result of low deposition rate and/or stronger erosion, as elsewhere in Quaternary environments.

The most important implication of this stage of the research is identifying that enhanced calcareousness in the Late Pleistocene strong red palaeosol is rather of post-pedogenic origin, unrelated to the major phase(s) of the Habonim pedocomplex development. The impressive macrorhizoliths developed here not earlier than the final stage of palaeosol development. Hence taxonomic identification of the major polygenetic palaeosol as typical Hamra or transitional soil type from Hamra to Husmas is justified.

Once again refined genesis and stadiality of palaeosols seem more effective as an explanation of pedogeomorphic *variations* between the Atlit site at AJU and Habonim than previous interpretations of their different chrono-stratigraphical position. However, we are fully aware that additional sequences are needed to confirm or disprove the interpretation presented for the Carmel coast.

Acknowledgments This research was supported by the Research Authority of the University of Haifa. The author thanks Dr Tatyana S. Gendler (United Institute of Physics of the Earth, RAS, Moscow, Russia) for magnetic measurements, Prof. Amotz Dafni for help in coastal plants identification, and Prof. Avraham Ronen for persistent collaboration.

References

Alonso-Zarza, AM, Genise, JF, Cabrera, MC, Mangas, J, Martín-Pérez, A, Valdeolmillos, A, Dorado-Valiño, M, 2008: "Megarhizoliths in Pleistocene Aeolian Deposits from Gran Canaria (Spain): Ichnological and palaeo environmental significance", in: *Palaeogeogr. Palaeoclimatol. Palaeoecol*, 265: 39–51.

Bar-Yosef, O, Belfer-Cohen, A, 1989: "The Origins of Sedentism and Farming Communities in the Levant", in: *J. World Prehist*, 3, 4: 447–498.

Czernik, J, Goslar, T, 2001: "Preparation of graphite targets in the Gliwice Radiocarbon Laboratory for AMS 14C dating", in: *Radiocarbon*, 43, 2A: 283–291.

Dan, J, Fine, P, Lavee, H, 2007: *Soils of the land of Israel, The "ERETZ" Series* (Geographic Research & Publication (in Hebrew, with English summary)).

Dan, J, Koyumdjiski, H (Eds.), 1979: *The classification of Israel soils.* [The Volcani Center, Bet Dagan (in Hebrew, with English summary)].

Dan, J, Yaalon, DH, Koyumdjisky, H, 1969: "Catenary Soil Relationships in Israel, the Netanya Catena on Coastal Dunes of the Sharon", in: *Geoderma*, 2: 95–120.

Danin, A, Yaalon, DH, 1982: "Silt Plus Clay Sedimentation and decalcification During Plant Succession in Sands of the Mediterranean Coastal Plain of Israel", in: *Isr J Earth Sci*, 31: 101–109.

Evans, ME, Heller, F, 2003: *Environmental Magnetism. Principles and Applications of Enviromagnetics* (USA: Academic Press).
Farrand, WR, Ronen, A, 1974: "Observations on the Kurkar–Hamra Succession on the Carmel Coastal Plain",in: *Tel-Aviv*,1: 45–54.
Fedoroff, N, 1997: "Clay Illuviation in Red Mediterranean Soils", in: *Catena*, 28: 171–189.
Fedoroff, N, Courty, M-A, Guo, Z, 2010: Palaeosoils and Relict Soils. In: Stoops, G, Marcelino, V, Mees, F (Eds.), *Interpretation of Micromorphological Features of Soils and Regoliths*, (Elsevier Science): 623–662.
Flora of Israel Online 2003–2006 (retrieved on the Internet 15.1.11. http://flora.huji.ac.il/browse.asp?).
Frechen, M, Neber, A, Tsatskin, A, Boenigk, W, Ronen, A, 2004: "Chronology of Pleistocene Sedimentary Cycles in the Carmel Coastal Plain of Israel", in: *Quatern. Int*, 121: 41–52.
Frumkin, A, Stein, M, 2004: "The Sahara-East Mediterranean Dust and Climate Connection Revealed by Strontium And Uranium Isotopes in a Jerusalem Speleothem", in: *Earth Planetary Sci. Let*, 217: 451–464.
Gile, LH, Petterson, FF, Grossman, RB, 1966: "Morphological and Genetic Sequences of Carbonate Accumulation in Desert Soils", in: *Soil Sci*, 101: 347–360.
Golik, A, 2000: Sediment Transport Pattern Along the Israeli Coast as Indicated by Sand Grain Size. National Institution of Oceanography, Report H 11/2000.
Gvirtzman, G, Netser, M, Katsav, E, 1998: "Late Glacial to Holocene Kurkar Ridges, Hamra Soils, and Dune Fields in the Coastal Belt of Central Israel", in: *Isr J. Earth Sci*, 47: 29–46.
Gvirtzman, G, Wieder, M, 2001: "Climate of the Last 53,000 Years in the Eastern Mediterranean, Based on Soil-Sequence Stratigraphy in the Coastal Plain of Israel",in: *Quaternary Sci Rev*, 20: 1827–1849.
Hall, GK, Calvo, R, 2005: Digital shaded relief map of Israel (retrieved on the Internet 15.1.11. http://www.cybaes.org/archive/downloads/Hall2005/PlXI.pdf).
Holliday, VT, 2004: *Soil in Archaeological Research* (Oxford University Press).
Kapur, S, Karaman, C, Akca, E, Aydin, M, Dinc, U, Fitzpatrick, EA, Pagliai, M, Kalmar, D, Mermut, AR, 1997: "Similarities and Differences of the Spheroidal Microstructure in Vertisols from Turkey and Israel", in: *Catena*, 28: 297–311.
Khadkikar, AS, Chamyal, LS, Ramesh, R, 2000: "The Character and Genesis of Calcrete in Late Quaternary Alluvial Deposits, Gujarat, Western India, and its Bearing on the Interpretation of Ancient Climates, in: *Palaeogeogr. Palaeoclimatol. Palaeoecol*, 162: 239–261.
Kutiel, P, 2001: "Conservation and Management of the Mediterranean Coastal Sand Dunes in Israel, in: *J Coast Conservat*, 7: 183–192.
McLaren, S, Gardner, R, 2004: "Late Quaternary Vadose Carbonate Diagenesis in Coastal and Desert Dune and Beach Sands: Is There a Palaeo Climatic Signal?",in: *Earth Surf. Process Landforms*, 29: 1441–1458.
Neber, A, 2002: *Sedimentological properties of Quaternary deposits on the Central coastal plain, Israel*. PhD thesis, (Haifa, Israel: University of Haifa).
Paton, TR, Humphreys, GS, Mitchell, PB, 1995: *Soils—A New Global View* (New Haven and London: Yale University Press).
Porat, N, Wintle, AG, Ritte, M, 2004: "Mode and Timing of Kurkar and Hamra Formation, Central Coastal Plain, Israel", in: *Isr J Earth Sci*, 53: 13–25.
Priori, S, Costantini, EAC, Capezzuoli, E, Protano, G, Hilgers, A, Sauer, D, Sandrelli, F 2008: "Pedostratigraphy of Terra Rossa and Quaternary Geological Evolution of a Lacustrine Limestone Plateau in Central Italy", in: *J. Plant Nutr. Soil Sci*, 171: 509–523.
Ronen, A, 1977: Mousterian Sites in Red Loam in the Coastal Plain of Mount Carmel, in: *Erez Israel*, 13: 183–190.
Ronen, A, Tsatskin, A, Laukhin, SA, 1999: "Genesis and Age of the Mousterian Paleosols in the Carmel Coastal plain, Israel", in: Davies W, Charles R (Eds.): *Dorothy Garrod and the Progress of the Palaeolithic Studies In the Prehistoric Archaeology of the Near East and Europe*, (Oxford: Oxbow Books): 135–151.
Singer, A, 2007: *The Soils of Israel* (Berlin, NewYork: Springer).

Sivan, D, Porat, N, 2004: "Evidence from Luminescence for Late Pleistocene Formation of Calcareous Aeolianite (Kurkar) and Paleosol (Hamra) in the Carmel Coast, Israel", in: *Palaeogeogr. Palaeoclimatol. Palaeoecol*, 211: 95–106.

Soil Taxonomy, A Basic Classification for Making and Interpreting Soil Surveys, 2nd edition 1999: Agriculture Handbook 436. USDA, Natural Resources Conservation Service, Washington.

Stoops, G, 2003: *Guidelines for Analysis and Description of Soil and Regolith Thin Sections* (USA: Soil Science Society of America).

Tsatskin, A, Ronen, A, 1999: "Micromorphology of Mousterian Paleosol in Aeolianites at the Site of Habonim, Israel",in: *Catena*, 34: 365–384.

Tsatskin, A, Gendler, TS, Heller, F, Dekman, I, Frey, GL, 2009: "Towards Understanding Paleosols in Southern Levantine Eolianites: Integration of Micromorphology, Environmental Magnetism and Mineralogy", in: *J Mt Sci*, 6: 113–124.

Tsoar, H, 2000: "Geomorphology and Paleogeography of Sand Dunes that have Formed the Kurkar Ridge in the Coastal Plain of Israel", in: *Isr. J. Earth Sci*, 49: 189–196.

Wieder, M, Gvirtzman, G, 1999: "Micromorphological Indications on the Nature of the Late Quaternary Paleosols in the Southern Coastal Plain of Israel", in: *Catena*, 35: 219–237.

Wieder, M, Yaalon, DH, 1974: "Effect of Matrix Composition on Carbonate Nodule Crystallization", in: *Geoderma*, 11: 95–121.

Wieder, M, Yaalon, DH, 1982: "Micromorphological Fabrics and Developmental Stages of Carbonates Nodular Forms Related to Soil Characteristics", in: *Geoderma*, 28: 203–220.

WRB (World Reference Base for Soil Resources), 1998: IUSS, ISRIC and FAO, Rome.

Yaalon, DH, 1967: "Factors Affecting the Lithification of Eolianite and Interpretation of its Environmental Significance in the Coastal Plain of Israel", in: *J Sediment Petrol*, 37: 1189–1199.

Yaalon, DH, 1997: "Soils in the Mediterranean Region: What Makes Them Different?, in: *Catena*, 28: 157–169.

Permissions and Credits

The editors and authors are grateful to the following copyright holders, publishers, authors, and photographers who have granted permission to use copyrighted material.

The cover photo was taken from: *A Mediterranean Soil Landscape in the Southern Coast of Anatolia, Ancient Kalantos* (Kaledran), (archive of Department of Archaeometry, University of Çukurova).

In Chap. 3 by Claudio Zucca, Riccardo Biancalani, Selim Kapur, Erhan Akça, Pandi Zdruli, Luca Montanarella and Freddy Nachtergaele on "Role of soil information in Land Degradation and Desertification mapping—A review" received the following permissions for the use of:

Table 3.5: Number of profiles in WISE3 by continent and their description status. **Source**: N.H. Batjes, 2009: Harmonized soil profile data for applications at global and continental scales: updates to the WISE database. *Soil Use and Management*, 25: 124–127. The permission to use this copyrighted table was granted by Wiley and Sons on 25 October 2012 with Licence Number 3015960915671.

In Chap. 5 by Ahmet R. Mermut on "Carbon Cycle and Sequestration in Terrestrial Ecosystems with Specific Reference to Mediterranean and Boreal Regions" used copyrighted material published by the Centre for Land and biological Resources Research, Agriculture and Agri-food Canada, Minister of Supply and Services Canada to use copyrighted material by D. F. Acton and L. J. Gregorich, 1995: *The health of our soils, towards the sustainable agriculture in Canada*. Agriculture and Agri-Food Canada, Research Branch. Center for Land and Biological Resources Publ. 1906/E that is in the public domain:

Table 5.3: Organic matter at two depths after eighteen years of various tillage treatments of a soil from Ontario, Canada under corn. Source: Acton and Gregorich (1995). Permission was granted.

Table 5.4: Organic matter present in the surface horizon (20 cm) cropped with corn and under different management schemes. Source: Acton and Gregorich (1995). Permission was granted.

Figure 5.2: Effects of no-till on soil organic carbon within the top 15 cm of the soil. Source: Acton and Gregorich (1995). Permission was granted.

S. Kapur and S. Erşahin (eds.), *Soil Security for Ecosystem Management*,
SpringerBriefs in Environment, Security, Development and Peace 8,
DOI: 10.1007/978-3-319-00699-4, © The Author(s) 2014

Figure 5.3: Soil organic matter levels as predicted by computer simulation models under three different management systems. Source: Acton and Gregorich (1995). Permission was granted.

Figure 5.4: Changes in soil organic matter over fifty years, with and without manure, in Alberta, Canada. Source: Acton and Gregorich (1995). Permission was granted.

In Chap. 7 by *Alexander Tsatskin on* "Polygenic red calcic soils in coastal Middle Palaeolithic environments, Israel: taxonomy and pedosedimentary reconstructions" the author used material from previous research, in which Dr. Frechen, Dr. Neber, and Prof. Boenigk all participated that took place more than ten years ago and dealt with stratigraphy and dating of eoalinites.

About the Authors

Riccardo Biancalani, born on 17 September 1960 in Florence/Italy. International consultant and technical advisor at FAO on land degradation assessment and natural resources management, with experience in Europe, Africa, Asia and South America.
Address: FAO, Rome, Italy.
Email: Riccardo.Biancalani@fao.org
Website: www.fao.org

Winfried E.H. Blum, born on 15 June 1941 in Freiburg, Germany. Professor emeritus at the Institute of Soil Research, University of Natural Resources and Life Sciences (BOKU), Vienna, Austria, co-editor or member of the editorial boards of 21 scientific journals, more than 550 scientific publications in nine languages in the fields of soil chemistry and mineralogy, land use, soil and environmental protection, and soil and society.
Address: Institute for Soil Research, BOKU, Vienna, Austria
Email: Winfried.blum@boku.ac.at
Website: www.boku.ac.at

Pierluigi Bucelli was senior researcher at CRA-ABP (Centro di ricerca per l'agrobiologia e la pedologia di Firenze); he retired in 2012. Over thirty years' experience in the quality of agricultural crops, with a particular interest in viticulture and oenology. Agronomist and oenologist, he is a member of the Italian Academy of Vine and Wine. He is author of over 100 scientific publications in national and international journals and books on grapevine and wine quality, soil survey for quality crops (vine, olive tree), and soil hydrology. He has specialist interests in soil management, environmental and grapevine ecophysiology, soil quality and viticultural and oenological results, and soil functional factors for viticultural and wine quality.
Address: CRA-ABP Agrobiology and Pedology Research Centre, Piazza M. d'Azeglio 30, 50121, Firenze, Italy
Email: pierluigi.bucelli@gmail.com
Website: http://abp.entecra.it/

S. Kapur and S. Erşahin (eds.), *Soil Security for Ecosystem Management*,
SpringerBriefs in Environment, Security, Development and Peace 8,
DOI: 10.1007/978-3-319-00699-4,

Edoardo A. C. Costantini is research director at the CRA-ABP Centro di ricerca per l'agrobiologia e la pedologia of Florence, where he leads the national soil database (<http://www.soilmaps.it/>). He was a lecturer in Pedology and Geopedology in the Department of Earth Sciences of the University of Siena from 1999 to 2008. He is the secretary of the European Society for Soil Conservation, president of the Commission on Pedology of the Italian Soil Science Society, past president of the IUSS Commission on Palaeopedology, and soil national expert for the Italian Ministry of Agriculture, as well as a member of the EU soil Working Group for the INSPIRE directive. Current research interests focus on the impact of climate change and management on soil characteristics and qualities, use of proximal sensors in soil survey, soil functional factors for precision viticulture and wine quality, and soil heritage. He has published more than 200 papers (45 ISI) and books regarding soil survey and land evaluation for quality crops (vines, olive trees), palaeopedology, soil genesis, soil heritage, soil hydrology, soil geodatabases, land degradation, and desertification.
Address: CRA-ABP Agrobiology and Pedology Research Centre, Piazza M. d'Azeglio 30, 50121 Firenze, Italy
Email: Edoardo.costantini@entecra.it
Website: http://abp.entecra.it/

Luca Montanarella has been working since 1992 as scientific officer in the European Commission. Leading the Soil Data and Information Systems (SOIL Action) activities of the Joint Research Centre in support to the EU Thematic Strategy for Soil Protection and numerous other soil related policies, like the Common Agricultural Policy (CAP), the UNCCD, UNFCCC, CBD. Responsible of the European Soil Data Centre (ESDAC), the European Soil Information System (EUSIS) and the European Soil Bureau Network (ESBN). Recently in charge of supporting the establishment of the Global Soil Partnership at FAO. More then 200 publications, books and reports. Numerous awards and memberships.
Address: European Commission, DG JRC, Ispra, Italy.
Email: luca.montanarella@jrc.it
Website: http://ies.jrc.ec.europa.eu

Ahmet Mermut, Professor at Saskatchewan (Canada) and Harran Universities (Turkey), born 1944 in Bitlis, Turkey. He works in the Soil Science department of the Harran University, Şanlıurfa, Turkey. He has chaired Division 1 of the International Union of Soil Sciences (IUSS), and currently chairs the European Confederation of Soil Science Societies (ECSSS). His worldwide studies are concerned with soil micromorphology and mineralogy, soil genesis, land management, land degradation, and conservation management. He has published many papers of high impact in international journals.
Address: Harran University, Department of Soil Science, Şanlıurfa, Turkey.
Email: <a.mermut@sask.ca>.
Website: <www.harran.edu.tr>

Freddy Nachtergaele, born 26 July 1949 in Ghent, Belgium. Retired senior officer land resources with the Food and Agriculture Organization (FAO). Coordinator of the Land Degradation Assessment in Drylands (LADA) project and co author of the Global Agro-ecological Zones Assessment for Agriculture (GAEZ) and the Harmonized World Soil Database (HWSD). More than 100 publications in the field of soil chemistry, soil classification, land evaluation, land degradation assessment and land use planning. Extensive field experience in Africa and Southeast Asia.
Address: FAO, Rome, Italy.
Email: freddy_nachtergaele@hotmail.it,
Website: www.fao.org

Franco Previtali, Professor at Milan Bicocca University. Born in 1942 in Italy, he is a former vice-president of the Italian Soil Science Society. His main fields are geopedology and interdisciplinary research within earth and soil sciences. He was one of the organizers of the First International Meeting on Land Degradation held in Adana, Turkey at 1996. His contribution to the science of land degradation has great merit in the field of Earth Sciences.
Address: Milan Bicocca University, Piazza della Scienza, 1-20126 Milan, Italy
Email: franco.previtali@unimib.it
Website: http://www.disat.unimib.it/ita/corso/CV-DOCENTI/CV-PREVITALI%20 FRANCO.pdf

Rattan Lal is a Distinguished Professor at the State University College of Food, Agricultural and Environmental Science, Columbus, Ohio. He is the recipient of the prestigious Norman E. Borlaug Award (2005) and the von Liebig Award (2006) for his contributions to and research in the sustainable management of soil and natural resources. His service to professional organizations includes Editor-in-Chief of the Encyclopedia of Soil Science, Co-Editor-in-Chief of Soil and Tillage Research, and Past-President of the Soil Science Society of America. He is a fellow of the Soil Science Society of America, the American Society of Agronomy, the Third World Academy of Sciences, and the American Association for Advancement of Science. He is a member of the US National Committee of Soil Science, and was a lead author on the Intergovernmental Panel on Climate Change and the UN Millennium Assessment. He has authored and edited over 1,200 journal articles.
Address: The Ohio State University College of Food, Agricultural and Environmental Science, 422D Kottman Hall, 2021 Coffey Road, Columbus, OH 43210
Email: lal.1@osu.edu
Website: http://senr.osu.edu/facview.asp?id=382

Alexander Tsatskin, Associate Professor, born 12 February 1952 in Moscow, USSR. He works in the Archaeology department of the University of Haifa, Israel. He co-chaired the International Union of Soil Science (IUSS) Soil Morphology

commission and his studies are focused on palaeopedological features, integration of micromorphology, and environmental magnetism and mineralogy, and are based on interdisciplinary approaches. He has published more than 100 papers on soil science.
Address: Zinman Institute of Archaeology, University of Haifa, Haifa 31905, Israel
Email: tsatskin@research.haifa.ac.il
Website:
http://archlgy.haifa.ac.il/index.php/en/faculty/43-prof-alexander-tsatskin.html

Pandi Zdruli, born in Berat, Albania on 23 December 1957. Professor of Soil Science and Natural Resources at the Mediterranean Agronomic Institute of Bari in Italy, one of the four affiliates of the International Centre for Advanced Mediterranean Agronomic Studies (CIHEAM). Author of 63 scientific papers and technical reports and member of the editorial boards of nine journals dealing with soil science, environmental issues and land degradation desertification research.
Address: Mediterranean Agronomic Institute of Bari, Land and Water Resources Management Department, Via Ceglie 9,7001, Valenzano (BA), Italy
Email: pandi@iamb.it
Website: http://www.iamb.it

Claudio Zucca, born 1 September 1972 in Brescia, Italy. Researcher in Pedology at the Department of Territorial Engineering of Sassari University, Italy, since 1999, in the framework of its collaboration with the NRD (Desertification Research Group). He has been involved in several international projects in the field of land degradation and desertification, his scientific interest being mainly focused on soil/land evaluation and on the integrated assessment of soil degradation and restoration in drylands, ranging from micromorphology to soil–landscape relationships. He has more than 50 scientific publications on soil science.
Address: Ricercatore in Pedologia, Dipartimento di Agraria, Università Degli Studi di Sassari, Viale Italia 39, 07100 Sassari, Italy
Email: clzucca@uniss.it
Website: http://80.24.165.149/drupal/?q=node/18

The Editors Other Books Published by Springer

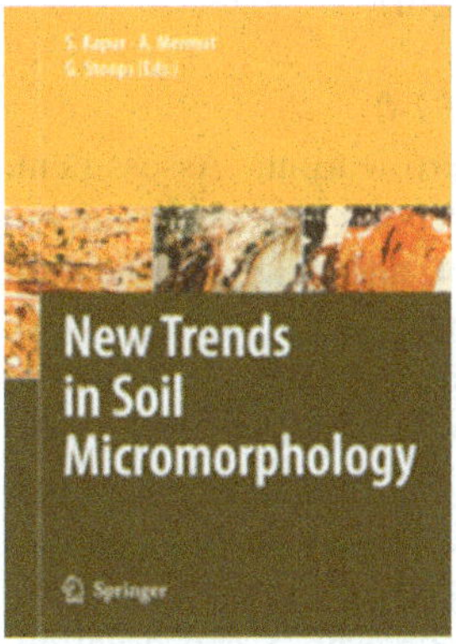

New Trends in Soil Micromorphology
Kapur, Selim; Mermut, Ahmet R.; Stoops, Georgess (Eds.)
2008, XIV, 276 p. 160 illus.
Hardcover ISBN 978-3-540-79133-1
Ebook ISBN 978-3-540-79134-8
The book contains state of the art new research results in micromorphology as well as other disciplines of soil science. It provides very useful up-to-date information for researchers, educators, graduate students interested in microscopic and submicroscopic studies of soils and sediments. In the past, micromorphology has been considered almost solely as a descriptive and interpretative branch of science. Attempts are now made to obtain quantitative data. There has been much progress in applying soil micromorphology in Quaternary geology, in particular identifying and characterizing palaeosols. The new areas for soil micromorphology are soil ecology, materials sciences and archaeology.

Sustainable Land Management
Learning from the Past for the Future
Kapur, Selim; Eswaran, Hari; Blum, Winfried E.H. (Eds.)
2011, XX, 400 p. 206 illus. Hardcover
ISBN 978-3-642-14781-4
Soil quality is threatened by many human-induced activities, but can also be improved by good land management. In the relatively short history of mankind on earth, the landscape and soils of the world have been drastically modified from their "natural "state. Landscapes altered by man's activities are termed "Anthroscapes" which are inextricably linked to

culture and history. The challenges for today's scientists are to devise and implement sustainable land management strategies in order to preserve the land for the benefit of future generations. This book is a valuable compendium of the research experiences so far gained in studies of the context and concept of the "Anthroscape" and highlights the potential future contributions of such research to sustainable development.

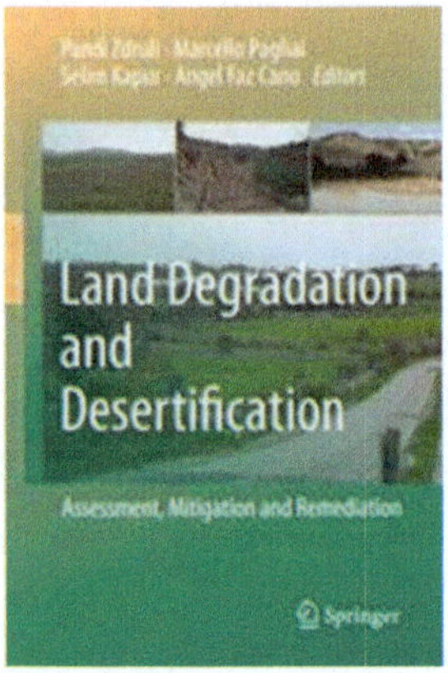

Land Degradation and Desertification:
Assessment, Mitigation and Remediation
Pandi Zdruli, Marcello Pagliai, Selim Kapur and Angel Faz Cano (Eds.)
2010,
ISBN: 978-90-481-8656-3 (Print)
978-90-481-8657-0 (Online)
DOI: 10.1007/978-90-481-8657-0
Land Degradation and Desertification: Assessment, Mitigation, and Remediation reports research results in sustainable land management and land degradation status and mitigation in 36 countries around the world. It includes background papers with continental and international perspectives dealing with land degradation and desertification studies. The book assembles various topics of interest for a large audience. They include carbon sequestration and stocks, modern techniques to trace the trends of land degradation, traditional and modern approaches of resource-base conservation, soil fertility management, reforestation, rangeland rehabilitation, land use planning, GIS techniques in desertification risk cartography, participatory ecosystem management, policy analyses and possible plans for action. Various climatic domains in Africa, Asia, Europe and The Americas are covered. The book will be of interest to a variety of environmental scientists, agronomists, national and international policy makers and a number of organizations dealing with sustainable management of natural resources.

About the Editors

Selim Kapur, born in 1946 in Ankara, Turkey. Professor of Soil Science at the Faculty of Agriculture and Chairman of the Dept. of Archaeometry at the University of Çukurova, Adana, Turkey. He is the Scientific Comm. Member of the European Soil Bureau Network of the EU-JRC-IES in Ispra, Milan, Italy. He is the earlier Secretary of the International Working Group of Land Degradation and Desertification. Has acted as the Science and Technology Correspondant of Turkey for the UNCCD. He organized the First International Meeting on Land Degradation and Desertification in 1996 in Adana, Turkey.

He led projects supported by NATO, EU, JSPS, TUBITAK, Mitsui Environmental Fund, FAO and GEF related to soil protection and environmental issues. He authored numerous papers in cited and authored/edited books concerning his major fields of study on Land degradation and desertification, archaeometry and soil micromorphology/soil mineralogy. Among his major publications are: Kapur, S., Karaman, C., E.Akça., Aydın, M., Dinç, U., FitzPatrick, E.A., Pagliai, M., Kalmar, D. and Mermut, A.R., 1997. Similarities and Differences Spheroidal Microstructure in Vertisol from Turkey and Israel. CATENA, Vol: 28 No: 3–4: 297–311; Cangir, C., Kapur, S., Boyraz, D., Akça, E., and Eswaran, H. 2000. An Assessment of Land Resource Consumption in Relation to Land Degradation Turkey. Journal of Soil and Water Conservation. 253–259; Zucca, C., Vignozzi, N., Madrau, S., Dingil, M., Previtali, F. And Kapur, S. 2013. Shape and intra-porosity of topsoil aggregates under maquis and pasture in the Mediterranean region. Journal of Plant Nutrition and Soil Science, Elsevier, Wiley. (İn Press); Eswaran, H., Kapur, S., Akca, E., Reich, P., Mahmoodi, S., and Vearasilp, T. 2005. Anthroscapes: A landscape unit for assessment of human impact on land systems. In: J.E.Yang, T.M. Sa, and J.J. Kim (Eds.) Application of the emerging soil research to the conservation of Agricultural Ecosystems. Publ.: The Korean Society of Soil Science and Fertilizers, Seoul, Korea; pp. 175–192; Kapur, S., Ryan, J., Akça, E., Çelik, İ., Pagliai, M., Tülün, Y. 2007. Influence of

S. Kapur and S. Erşahin (eds.), *Soil Security for Ecosystem Management*,
SpringerBriefs in Environment, Security, Development and Peace 8,
DOI: 10.1007/978-3-319-00699-4, © The Author(s) 2014

Mediterranean Cereal-based Rotations on Soil Micromorphological Characteristics. GEODERMA 142. 318–324.

Address: University of Çukurova, Department of Soil Science and Archaeometry, 01330, Adana, Turkey.

Email: <kapurs@gmail.com>.

Website: <www.cu.edu.tr>.

Sabit Erşahin was born on 4 December 1963 in Artvin, Turkey, and received his BS and MS degrees from the Department of Soil Science, University of Çukurova. Obtained his Ph.D. from Washington State University on 'The Solute Transport in Sloping Layered Soils' under the supervision of Prof. R.I. Papendick in 1996. His major field of study has been soil physics, soil physico-chemistry, hydropedology, modelling, soil management, and alternative agriculture. He presently serves as the vice-rector of the Karatekin University in Çankırı, Turkey.

Among his major publications are: Ersahin, S. 2001. Assessment of Spatial Variablity in Nitrate Leaching to Reduce Nitrogen Fertilizers Impact on Water Quality. Agricultural Water Management. 48/3, 179–189; Ersahin, S. 2003. Comparing ordinary kriging and cokriging to estimate infiltration rate. Soil Sci., Soc., Am., J. 67: 1848–1855; Ersahin S., and A. R. Brohi. 2006. Spatial variation of soil water content in topsoil and subsoil of a Typic Usitfluvent. Agricultural Water Management. 83:79–86; Ersahin, S. H. Gunal, T. Kutlu, B. Yetgin, and S. Coban. 2006. Estimating specific surface area and cation exchange capacity in soils with fractal dimension of particle size distribution. Geoderma, 136: 588–597; Kurunç, A, Ersahin, S., Uz, BY., Sonmez, NK., Uz, I, Kamah, H, Bacalan GE, Emekli Y. 2011. Identification of nitrate leaching hot spots in a large area with contrasting soil texture and management. Agricultural Water Management: 98(6), 1013–1019. DOI: 10.1016/j.agwat.2011.01.010

Address: Çankırı Karatekin University, Department of Forestry Engineering, Çankiri, Turkey.

Email: <acapsu@gmail.com>.

Website: <sabitersahin.com/index.htm>.

Erhan Akça, born in 11 March 1969 in adana, turkey. Assoc.Prof. Dr. at Technical Programmes, Adıyaman University, Adıyaman, Turkey. His profession is on soil mineralogy, archaeometry and natural resource management. He wrote and contributed to more than 100 scientific publications (papers, chapters, books) on soil science, archaeometry, land use and desertification. He participated to projects supported by NATO, EU, JSPS, TUBITAK, Mitsui Environmental Fund, FAO and GEF. He is currently scientific board member of Ministry of Forestry and Water Works, General Directorate of Combating to Desertification and Erosion Control of Turkey.

Among his major publications are: Kapur, S., Ryan, J., Akça, E., Çelik, İ., Pagliai, M., Tülün, Y. 2007. Influence of Mediterranean Cereal-based Rotations on Soil Micromorphological Characteristics. GEODERMA 142. 318–324; Akça, E., Çimrin, K.M., Ryan, J., Nagano, T., Topaksu, M., Kapur, S. 2008. Differentiating the natural and man-made terraces of Lake Van, Eastern Anatolia, utilizing earth science methods. Lakes & Reservoirs: Research and Management, 13, (1). (March 2008), 83–93; Ryan, J., Kapur, S. and Akça, E. 2009. Application of Soil Analyses as Markers to Characterise a Middle Eastern Chalcolithic—Late Bronze Age Mound. Turkish Academy of Sciences Journal of Archaeology (TUBA-AR). 12: 65–76; Akça, E., Kapur, S., Tanaka, Y., Kaya, Z., Bedestenci, H.Ç. Yaktı, S. 2010. Afforestation Effect on Soil Quality of Sand Dunes. Polish Journal of Environmental Studies. Vol. 19, No. 6, 1109–1116; Akça, E., Fujikara, R. and Sabbağ, Ç. 2013. Atatürk Dam resettlement process: increased disparity resulting from insufficient financial compensation. International Journal of Water Resources Development Vol. 29, No. 1, 101–108.

Address: Technical Programmes, Adıyaman University, Adıyaman, Turkey
Email: *erakca@gmail.com*
Website: http://www.adiyaman.edu.tr/abys/index.php?git=a&a_id=80

About the Book

This book contains invited and peer-reviewed chapters for the first issue of the Subseries on *Mediterranean Soil Ecosystems* (MSE), a book publication of the *Soil Science Society of Turkey* (SSST). This first volume contains seven chapters on "Managing Terrestrial Carbon in a Changing Climate" by Lal Rattan (USA), on "Land Degradation and Security Linkages in the Mediterranean Region" by Winfried E. H. Blum (Germany/Austria), on "Role of Soil Information in Land Degradation and Desertification Mapping: A Review" by Claudio Zucca, Riccardo Biancalani (Italy), Selim Kapur, Erhan Akça (Turkey), Pandi Zdruli, Luca Montanarella (Italy), and Freddy Nachtergaele (Belgium/FAO), on "Pedoenvironments of the Mediterranean Countries. Resources and Threats" by Franco Previtali (Italy), on "Carbon Cycle and Sequestration in Terrestrial Ecosystems with Specific Reference to Mediterranean and Boreal Regions" by Ahmet R. Mermut (Canada/Turkey), on "Soil and Terroir" by Edoardo Costantini and Pierluigi Bucelli (Italy), and on "Polygenic Red Calcic Soils in Coastal Middle Palaeolithic Environments, Israel: Taxonomy and Pedosedimentary Reconstructions" by Alexander Tsatskin (Israel). These renowned authors are experts on Mediterranean land use and traditional and/or traditional renovated technologies, on land management approaches adapted or sought for the sustainable production systems in the Mediterranean Region, the Integrated context of sustainable C-management, the development of the interdisciplinary soil ecosystem management approach, integrated inferences in developing soil quality attributes from the microscope to the field, land/soil degradation and technologies in a global scale, and ultimately soil change bound to the impacts of the foreseen future climate change in the Mediterranean region.

S. Kapur and S. Erşahin (eds.), *Soil Security for Ecosystem Management*, SpringerBriefs in Environment, Security, Development and Peace 8,
DOI: 10.1007/978-3-319-00699-4, © The Author(s) 2014

The manufacturer's authorised representative in the EU is Springer Nature Customer Service Centre GmbH, Europaplatz 3, 69115 Heidelberg, Germany. If you have any concerns regarding our products, please contact ProductSafety@springernature.com

Printed and bound by CPI Group (UK) Ltd, Croydon, CR0 4YY
30/09/2026
02225643-0019